FLORA ZAMBESIACA

Flora terrarum Zambesii aquis conjunctarum

VOLUME EIGHT: PART ONE

FLORA ZAMBESIACA

MOZAMBIQUE

MALAWI, ZAMBIA, ZIMBABWE

BOTSWANA

VOLUME EIGHT: PART ONE

Edited by

E. LAUNERT

on behalf of the Editorial Board:

E. A. BELL
Royal Botanic Gardens, Kew

E. LAUNERT
British Museum (Natural History)

I. MOREIRA
Centro de Botânica, Instituto de Investigação
Científica Tropical, Lisboa

Published by the Managing Committee on behalf of
the contributors to Flora Zambesiaca
1987

Photoset by Rowland Phototypesetting Limited
Bury St Edmunds, Suffolk
Printed in Great Britain by
St Edmundsbury Press Limited
Bury St Edmunds, Suffolk

ISBN 0 9507682 2 7

CONTENTS

LIST OF FAMILIES INCLUDED IN
VOLUME VIII, PART 1

CONVOLVULACEAE

By Maria Leonor Gonçalves

Herbs or shrubs, rarely small trees, frequently twining or prostrate, less often erect, provided with diverse sorts of glandular and eglandular hairs, besides simple, 2-armed or stellate hairs. Leaves alternate, exstipulate, usually simple, entire or toothed to often lobed. Flowers often large and showy, usually bracteate, axillary or terminal, solitary or in various inflorescences, almost always regular, bisexual save in a very few small genera, 5-merous as to the calyx, corolla and androecium (4-merous in *Hildebrandtia*). Sepals imbricate, sometimes unequal, generally free or connate at the base, often accrescent. Corolla sympetalous, variable but often funnel-shaped or salver-shaped, entire or 4–5-lobed, induplicate-valvate or contorted. Stamens as many as and alternate with lobes or connate members of the corolla, inserted in the corolla tube; filaments often unequal; anthers tetrasporangiate or dithecal, opening by longitudinal slits; pollen smooth or spinulose. Ovary superior, entire or 2–4-lobed, of 2(3–5) carpels united, (1)4(3–5)-locular, usually with an annular nectary-disk around the base; styles 1–2 (3) mostly terminal; stigmas 1–4 variously shaped; ovules 2 per carpel (rarely many in *Humbertia*), basal or basal-axile, erect, anatropous. Fruit usually dry, a loculicidal (or sometimes irregularly dehiscent) capsule, or less often indehiscent and baccate or nut-like. Seeds 1–4 (rarely 6 or 10) with endosperm; embryo large, straight or curved, with 2 plicate, often bifid cotyledons, embedded in a hard, often cartilaginous endosperm.

This is a well-defined family; the genera are difficult to limit satisfactorily. It consists of about 50 genera and 1500 species, nearly all cosmopolitan in distribution, but best developed in tropical and subtropical regions. I have followed the classification used by van Ooststroom in Flora Malesiana, Ser. 1, 4: 388–512 (1953) and Verdcourt in Flora Tropical East Africa, Convolvulaceae (1963).

1. Ovary distinctly 2 or 5-lobed; style 2 (sometimes connate below), inserted between the lobes of the ovary; small prostrate herbs - - - - - - - 2
— Ovary not deeply lobed; style simple or, if styles 2, terminal; habits various - - 3
2. Ovary 2-lobed - - - - - - - - - - - **1. Dichondra**
— Ovary 4-lobed - - - - - - - - - - - **2. Falkia**
3. Styles 2, quite or almost separate - - - - - - - - - 4
— Styles 1 or 2, partly joined - - - - - - - - - - 7
4. Each style forked for about half its length - - - - - **3. Evolvulus**
— Styles not forked - - - - - - - - - - - 5
5. Large climbing shrub - - - - - - - - - - **5. Bonamia**
— Small erect or prostrate subshrubs - - - - - - - - 6
6. Stamens and styles included - - - - - - - - **4. Seddera**
— Stamens and styles exserted - - - - - - - - **6. Cressa**
7. Styles 2, jointed for about half their length - - - - - **5. Bonamia**
— Style 1 - - - - - - - - - - - - 8
8. Stigmas linear, about the same thickness as the styles, filiform - - **8. Convolvulus**
— Stigmas oblong to globose, rarely filiform - - - - - - 9
9. Indumentum conspicuously stellate-hairy - - - - **16. Astripomoea**
— Indumentum not stellate (except in *Merremia stellata* and *Ipomoea ephemera*) but sometimes with a few scattered branched hairs - - - - - 10
10. Stigmas oblong or nearly so, rarely filiform - - - - - - 11
— Stigmas globose or nearly so - - - - - - - - - 13
11. Leaves rounded at the base; 3 outer sepals largest and decurrent on to the peduncle; a marsh plant of the coast, with usually solitary flowers - - - - **10. Aniseia**
— Leaves cordate, hastate or cuneate at the base; sepals equal or unequal, not decurrent; plant of various habitats (not as above), with few-flowered or capitate inflorescences - - 12
12. Stigmas subglobose to filiform usually oblong or elliptic; corolla usually blue but sometimes white or pink, about 1 cm. long - - - - - **7. Jaquemontia**
— Stigmas ovate-oblong, complanate; corolla yellow, rarely white, usually with maroon or claret centre, 2–3·5 cm. long - - - - - - - **9. Hewittia**
13. Corolla narrow, urceolate-tubular, constricted below, distinctly though slightly curved, serrate at the apex - - - - - - - - - **14. Mina**
— Corolla not shaped as above - - - - - - - - - 14

9

14. Corolla urceolate, constricted at apex of tube · · · · · **13. Lepistemon**
— Corolla not distinctly urceolate · · · · · · · · · 15
15. Pollen smooth; corolla nearly always yellow or white with or without a red or purple
centre · · · · · · · 16
— Pollen spinous; corolla variously coloured, often purple · · · · · 17
16. Fruit a 4-valved capsule; bracts usually small (at most up to 12 mm. long)   **11. Merremia**
— Fruit a large capsule with circumscissile epicarp; bracts large (15–25 mm. long)
                            12. Operculina
17. Fruit indehiscent or pericarp opening irregularly · · · · · · 18
— Fruit a 3–10 valved capsule · · · · · · · **17. Ipomoea**
18. Fruit completely enclosed by the much enlarged calyx; pericarp thin, irregularly; seeds 4;
leaves more or less glabrous and densely covered with minute glands on the inferior surface
(appearing as black dots in dried specimens) · · · · **15. Stictocardia**
— Fruit usually not enclosed by the calyx; pericarp woody or leathery, indehiscent; seeds
usually 1; leaves usually densely pubescent on both surfaces · · · · · 19
19. Stamens inserted directly on the corolla tube; style not completely caducous in the fruit
 · · · · · · · · · · · **18. Turbina**
— Staemens inserted on scales situated towards the base of the corolla tube; style completely
caducous in the fruit · · · · · · · **19. Paralepistemon**

Argyreia nervosa (Burm. f.) Boj., a climber up to 10 m. high, densely whitish or fulvous tomentose, with large ovate or circular-cordate leaves and pink-purple c. 6 cm. long flowers, is native in India from Assam and Bengal to Belgaum and Mysore and cultivated as a garden plant in Mozambique.

1. DICHONDRA J. R. & G. Forst.

Dichondra J. R. & G. Forst., Char. Gen.: 39, t. 20 (1776).—Tharp. & M. C. Johnston in Brittonia **13**: 347 (1961).

Small prostrate perennial herbs, glabrous or softly hairy. Leaves simple, petiolate, cordate-circular or reniform, entire. Flowers small, solitary, axillary, pedicellate; bracteoles 2, minute, subulate. Sepals 5, more or less free, scarcely jointed at the base, subequal, ovate-spathulate, somewhat accrescent. Corolla widely campanulate, deeply 5-lobed, not longer than the calyx, hirsute outside. Stamens 5, shorter than the corolla; filaments filiform; anthers small; pollen smooth. Ovary deeply bi-lobed; each lobe with 2 ovules; styles 2, gynobasic and inserted between the lobes, short, filiform; stigmas capitate. Capsule bi-lobed; lobes subglobose, membranous, usually single-seeded, indehiscent or irregularly bivalved. Seeds subglobose, smooth.

A small genus of about five species, principally New World, with one species in the tropical and subtropical regions of both hemispheres.

Dichondra repens J. R. & G. Forst., Char. Gen.: 40, t. 20 (1775).—Engl. & Prantl, Pflanzenfam. ed. 1, **4**, 3a: 14 (1891).—Hall. f. in Engl., Bot. Jahrb. **18**: 82 (1893).-Hiern, Cat. Afr. Pl. Welw. **1**: 723 (1898).—Baker & Wright in Dyer, F. C. **4**: 83 (1904).—Baker & Rendle in F.T.A. **4**: 65 (1905).—Wild, Common Rhod. Weeds, fig. 93 (1955).—Dandy in F. W. Andr., Fl. Pl. Anglo-Egypt. Sudan **3**: 109 (1956).—Meeuse in Bothalia **6**: 657 (1958).—Heine in F.W.T.A. ed. 2, **2**: 338 (1963).—Verdc. in F.T.E.A., Convolvulaceae: 12, fig. 2 (1963).—Binns, H.C.L.M.: 39 (1968).—Ross, Fl. Natal: 294 (1972). TAB. 1. Type from New Zealand.

Var. **repens**

Procumbent herb, moderately pubescent, not densely appressed-pubescent nor sericeous. Stems slender, rooting at the nodes. Leaf lamina reniform to cordate-circular, 6–29 × 7–36 mm., broadly cordate at the base, broadly rounded or emarginate at the apex, appressed-hairy on inferior surface, usually glabrescent on the superior surface, more or less 7-nerved at the cordate base; petiole 8–70 mm. long usually appressed-hairy. Flowers on peduncles shorter than the petiole, 2–19 mm. long, hairy like the petioles. Calyx up to 3 mm. long, silky outside; sepals ovate-oblong to spathulate, obtuse, hairy on back and margins, prominently veined up to 2·5 × 1·5 mm. Corolla white or greenish, about as long as the calyx. Lobes of the capsule 1·5–2·5 mm. in diam., pilose. Seeds slightly trigonous, brown.

Zimbabwe. W: Masvingo, fl. & fr. 4.v.1962, *Drummond* 7938 (SRGH). C: Harare, Harris Rd., fl. & fr. 20.ii.1964, *Whitaker* in GHS 151397 (BM; SRGH). E: Mutare, 215 m., fl. & fr.

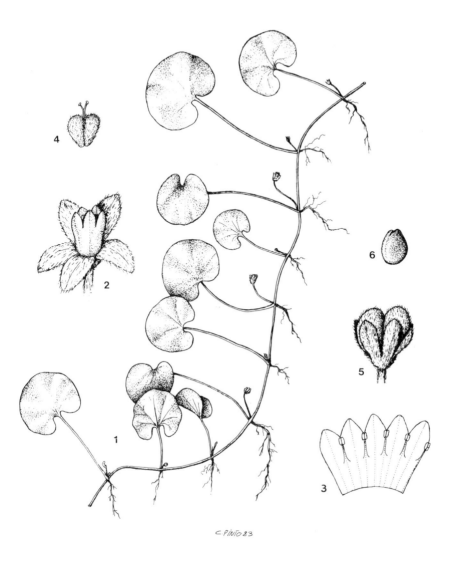

C PINTO 83

Tab. 1. DICHONDRA REPENS. 1, habit (×½); 2, flower (×5); 3, corolla opened to show stamens (×5); 6, seed (×5). All from *Chase* 4881.

24.iii.1953, *Chase* 4881 (BM; K; SRGH). S: Zimbabwe Ruins, fl. & fr. 17.iv.1946, *Greatrex* 14771 (SRGH). **Malawi**. N: Nyika Plateau, Kasaramba View, 2345 m., fl. & fr. 14.v.1970, *Brummitt* 10701 (BR; K; LISC; MAL; PRE; SRGH; UPS). **Mozambique**. GI: Gaza, Chirindzeni Forest, St. 1917–18, *Junod* 370 (LISC). M: Maputo, Macocololo, fl. & fr. 19.i.1898, *Schlechter* 12056 (PRE).

Widespread throughout the warmer regions of both hemispheres. Weed in grassland, lawns, paths and cultivated areas; 215–1500 m.

Var. *sericea* is densely appressed pubescent and sericeous and has discolourous leaves; it occurs in America and Australia.

2. FALKIA L. f.

Falkia L. f., Suppl. Pl.: 30 (1782).

Dwarf, prostrate perennial herbs. Leaves petiolate, oblong to oblong-lanceolate, entire. Flowers small, solitary, axillary, peduncled. Calyx shorter than the corolla tube. Sepals broad, ovate, forming a short tube, subequal, accrescent. Corolla funnel-shaped, plicate, shortly and broadly 5-lobed. Stamens 5, included; filaments linear, anthers oblong. Ovary deeply 4-lobed with an ovule in each lobe, hairy; styles 2, gynobasic, almost equalling the corolla tube; stigmas capitate. Capsule 4-lobed; lobes subglobose, single-seeded.

A genus with 3 species confined to Africa.

Falkia oblonga Bernh. ex Krauss in Flora **27**: 830 (1844).—Hall.f. in Engl., Bot. Jahrb. **18**: 84 (1893) & in Bull. Herb. Boiss. **7**: 41 (1899).—Baker & Wright in Dyer, F.C. **4**: 82 (1904).—Baker & Rendle in F.T.A. **4**: 65 (1905).—Meeuse in Bothalia **6**: 660 (1958). —Verdc. in F.T.E.A., Convolvulaceae: 14 (1963).—Roessler in Merxm., Prodr. FL. SW. Afr. **116**: 6 (1967).—Ross, Fl. Natal: 294 (1972). TAB. **2**. Type from Natal.

Var. **oblonga**

Rootstock sending stems creeping underground and appressed-pilose above. Leaves about twice as long to several times as long as broad; leaf lamina oblong to oblong-ovate 10–50 × 3–14 mm., rounded at the apex, truncate or slightly cuneate at the base, 5-nerved from the base, silvery silky, glabrescent above, appressed pilose beneath; petiole up to 80 mm. long. Flowers on pedicels 8–60 mm. long. Calyx turbinate; tube up to 3 mm. long; lobes oblong, acute, appressed-pilose outside. Corolla white to pale mauve up to 12 mm. long; lobes emarginate. Ovary densely pilose; styles about 8 mm. long. Capsule pilose. Seeds ovoid or globose.

Botswana. *N*: Ngamiland, 940 m., fl. & fr. 28.xii.1972, *Smith* 305 (K; SRGH). **Zambia.** S: Mumbwa, Kafue floodplain, S. of Namukumbo, fl. & fr. 9.ix.1964, *van Rensburg* 2961 (K; SRGH). **Mozambique.** GI: Inhambane, Banamana Salina, 20 km. SW. of Mabote, fl. & fr. x.1973, *Tinley* 2964 (K; LISC; SRGH). M: Maputo, fl. & fr. viii-xi, *Sim* 445 (PRE).
Also in Uganda and S. Africa. In damp grassland and saline sandy clays, up to 1000 m.

The var. *minor* C. H. Wright has almost sessile flowers and smaller leaves. It is recorded from Ethiopia and S. Africa.

3. EVOLVULUS L.

Evolvulus L., Sp. Pl., ed. **2**: 391 (1762) & Gen. Pl., ed. **6**: 152 (1764). *Volvulopsis* Roberty in Candollea **14**: 28 (1952).

Annual or perennial herbs or subshrubs, not twining. Leaves usually small, simple, entire, various, often sessile. Inflorescences usually axillary cymes or terminal and then spike-like. Flowers small; bracteoles small. Sepals small, subequal acute or obtuse, not accrescent. Corolla funnel-shaped or campanulate to subrotate or salver-shaped, entire or lobed. Stamens 5, inserted above the middle of the corolla tube, rarely near the base; filaments linear, glabrous; anthers oblong. Ovary ovoid or globose, 1–2-locular, each loculus 2-ovulate; styles 2, filiform, free from the base, each forked; stigmas 4, linear-terete or slightly clavate. Capsule avoid or globose, 2–4 valved, 1–2-locular. Seeds 4 or fewer by abortion, glabrous.

A genus of about 100 species, all American except two.

Corolla deeply lobed; capsule 1-locular; plant creeping prostrate -   - 1. *nummularius*
Corolla shallowly lobed; capsule 2-locular; plant more or less ascending -   2. *alsinoides*

1. **Evolvulus nummularius** (L.) L., Sp. Pl., ed. 2: 391 (1762).—Peter in Engl. & Prantl, Pflanzenfam. ed. 1, **4**, 3a: 19 (1891).—Hall. f. in Engl., Bot. Jahrb. **18**: 85 (1893). —Dammer in Engl., Pflanzenw. Ost.-Afr. **C**: 328 (1895).—Hiern, Cat. Afr. Pl. Welw. **1**: 723 (1898).—Baker & Rendle in F.T.A. **4**: 68 (1905).—van Ooststr. in Meded. Bot. Mus. Herb. Rijksuniv. Utrecht **14**: 114 (1934).—Dandy in F. W. Andr., Fl. Pl. Anglo-Egypt. Sudan **3**: 109 (1956).—Heine in F.W.T.A. ed. 2, **2**: 339 (1963).—Verdc. in F.T.E.A., Convolvulaceae: 16, fig. 4 (1963). Type from Jamaica.

Tab. 2. FALKIA OBLONGA. 1, habit (×½), from *Moura & al.* 296; 2, habit (×½) 3, flower opened (×2); 4, pistil (×3); 5, fruit (×2); 6, seed (×8), 2–6 from *van Rensburg* 2961.

14 CONVOLVULACEAE

Convolvulus nummularius L., Sp. Pl.: 157 (1753). Type as above.
Evolvulus dichondroides Oliv. in Trans. Linn. Soc., Bot. **29**: 117, t. 78b (1875). Type from Uganda.
Volvulopsis nummularius (L.) Roberty in Candollea **14**: 28 (1952). Type as above.

Perennial herb with the same habit as *Dichondra repens*. Stems prostrate, rooting at the nodes, pubescent. Leaf-lamina circular or circular-obovate, 5–20 mm. long and broad, apex very obtuse, base truncate to subcordate, glabrous or sparingly pubescent beneath; petiole 1–8 mm. long, pubescent, canaliculate above. Flowers solitary or rarely paired, axillary; peduncles up to 5 mm. long, recurving. Sepals ovate to ellipti-covate, 2·5–3 × 1·5 mm., more or less acute, pubescent or glabrescent with ciliate margins. Corolla white, subrotate, about twice as long as the calyx, deeply lobed; lobes obovate, 1·5–2·5 × 1–1·5 mm. Capsule globose 3–4 mm. in diam., unilocular, bivalved, 4(2)-seeded. Seeds brown to black, subglobose 1·5–2 mm. long, shining.

Zambia. N: Mporokoso, Mweru-Wa-Ntipa, Kanjiri, 1050 m., fl. & fr. 10.iv.1957, *Richards* 9102 (K). **Malawi**. S: Chikwawa, Lengwe Game Reserve, 100 m., fl. & fr. 9.iii.1970, *Brummitt & Hall-Martin* 8977 (EA; K; MAL; LISC; PRE; SRGH; UPS). **Mozambique**. MS: Beira, Gorongosa National Park, fl. & fr. iii.1972, *Tinley* 2499 (K; LISC; SRGH).
Tropical Africa and tropical America. In dry forest and dense thicket; 100–1050 m.

2. **Evolvulus alsinoides** (L.) L., Sp. Pl., ed. 2: 392 (1762).—Peters, Reise Mossamb. Bot. **1**: 246 (1861).—Peter in Engl. & Prantl, Pflanzenfam. ed. 1, **4**, 3a: 19 (1891).—Hall. f. in Engl., Bot. Jahrb. **18**: 85 (1893).—Dammer in Engl., Pflanzenw. Ost.-Afr. **C**: 328 (1895).—Hiern, Cat. Afr. Pl. Welw. **1**: 724(1898).—Hall. f. in Warb., Kunene-Samb.-Exped. Baum: 345 (1903).—Baker & Wright in Dyer F.C. 4: 79 (1904).—Baker & Rendle in F.T.A. **4**: 67 (1905).—Schinz in Gomes e Sousa, Pl. Menyharth.: 69 (1905).—Eyles in Trans. Roy. Soc. S. Afr. **5**: 452 (1916).—Fries, Wiss. Ergebn. Schwed. Rhod.-Kongo Exped. **1**: 268 (1916).—van Ooststr. in Meded. Bot. Mus. Herb. Rijksuniv. Utrecht **14**: 26 (1934). —Gomes e Sousa Pl. Menyarth. in Bol. Soc. Est. Mocamb. **32**: 86 (1936).—Wild, Guide Fl. Victoria Falls: 155 (1953).—Dandy in F. W. Andr., Fl. Pl. Anglo-Egypt. Sudan **3**: 109 (1956).—Meeuse in Bothalia **6**: 661 (1958).—Hiern, in F.W.T.A. ed. 2, **2**: 339 (1963). —Verdc. in F.T.E.A., Convolvulaceae: 18 (1963).—Roessler in Merxm., Prodr. Fl. SW. Afr. **116**: 6 (1967).—Binns, H.C.L.M.: 39 (1968).—Jacobsen in Kirkia **9**: 171 (1973). —Compton, Fl. Swaziland: 474 (1976). **3**. Type from Sri Lanka.
Convolvulus alsinoides L., Sp. Pl.: 157 (1753). Type as above.
Convolvulus linifolius L., Amoen. Acad. **4**: 306 (1759). Type from Senegal.
Evolvulus linifolius (L.) L., Sp. Pl., ed. 2: 392 (1762). Type as above.
Evolvulus azureus Schumach. & Thonn., Beskr. Guin. Pl.: 166 (1827). Type from Ghana.
Evolvulus fugacissimus A. Rich., Tent. Fl. Abyss. **2**: 75 (1851). Type from Ethiopia.

An extremely variable annual or perennial herb, thinly or sometimes rather densely covered with somewhat long patent silky hairs. Stems few to several, spreading or ascending, slender, up to 50 cm. long. Leaves subsessile; leaf lamina elliptic, ovate-oblong, lanceolate to linear-oblong, 5–45 × 1–15 mm., acute or rounded at both ends, distinctly mucronate, silky pilose on both surfaces; petiole up to 3 mm. long. Inflorescences axillary, 1-few-flowered; peduncle filiform, 5–50 mm. long, shorter to much longer than the leaves; bracts minute, lanceolate, up to 5 × 0·75 mm.; pedicesl filiform, short, up to 10 mm. long, spreading. Calyx densely silky or villous, sepals ovate-lanceolate, up to 5 × 1 mm. Corolla blue, rarely white, the folds paler beneath, broadly funnel-shaped, up to 8 mm. long and wide. Ovary 2-celled, each cell 2-ovuled, glabrous; styles 2, stigmas 4, long, terete or subclavate. Capsule globose, 3–4 mm. long, glabrous, 4-valved, 4-seeded. Seeds brown to black, ovoid, 1.7 mm. long, smooth, glabrous.

Botswana. N: c. 15 km. SW of Maun, c. 900 m., fl. & fr. 24.i.1972, *Biegel & Russell* 3775 (K; LISC; SRGH). SW: Ghanzi, farm 47, fl. & fr. 12.ii. 1969, *de Hoogh* 23 (SRGH). SE: Mochudi, fl. & fr. viii.1914, *Harbor* 14112 (PRE). **Zambia**. B: Senanga, Sioma, 1050 m., fl. & fr. 1.ii.1975, *Brummitt, Chisumpa & Polhill* 14196 (K). N: Mbala, Kasaba Sands, Lake Tanganyika, 780 m., fl. & fr.16.ii.1959, *Richards* 10991 (K; SRGH). W. Solwezi-Kafue, Musnama Fr., fl. & fr. 10.1.1962, *Holmes* 1435 (K). C: Lusaka, 1220 M., 21.iii.1961, *Best* 286 (K; SRGH). E: Chipata, 1100 m., fl. & fr. 4.i.1959, *Robson* 1030 (BM; K; LISC; SRGH). S: Machili, fl. & fr. 6.xii.1960, *Fanshawe* 5949 (K; SRGH). **Zimbabwe**. N: Chipuriro Distr., near Kanyemba Camp, c. 600 m., fl. & fr. 1.ii.1966, *Müller* 320 (K; SRGH). W: Hwange, 5.5 km. NW. of main Camp, nr. Impofu, fl. & fr. 17.xiii.1968, *Rushworth* 1364 (K; SRGH). C: Makoni, Inyazura, fl. & fr.

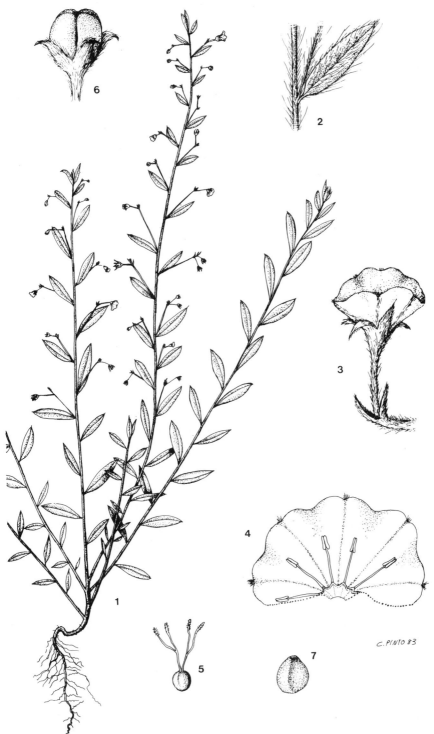

Tab. 3. EVOLVULUS ALSINOIDES. 1, habit ($\times\frac{1}{2}$); 2, portion of stem with leaf and peduncle ($\times$2); 3, flower ($\times$3); 4, corolla opened to show stamens ($\times$5); 5, pistil ($\times$5); 6, fruit ($\times$5); 7, seed ($\times$5). All from *Richards* 10991.

30.xii.1930, *Fries, Norlindh & Weimarck* 4015 (COI; K; SRGH). E: Mutare, Sports Grounds, 110 m., fl. & fr.15.iii.1960, *Chase* 7316 (BM; K; SRGH). S: Beit Bridge, Nulli Range, 730 m., fl. & fr. 10.i.1961, *Leach* 10669 (K; SRGH). **Malawi**. N: Mzimba, Mzuzu towards Mzambazi, fl. & fr. 25.iii.1972, *Pawek* 5111 (K; SRGH). C: Kasungu-Lundazi Rd., 1050 m., fl. & fr. 15.i.1959, *Robson* 1193 (BM; K; LISC; SRGH). S: Blantyre, fl. & fr. 29.i.1957, *Banda* 347 (BM; K; SRGH). **Mozambique** . N: Macondes, 6 km. from Nantulo towards Mueda, 300 m., fl. & fr. 10.v.1964, *Torre & Paiva* 11846 (LISC). T: Between Mutarara Velha and Sinjal, 42.8 km. from Mutarara Velha, fl. & fr. 18.vi.1949, *Barbosa & Carvalho* 3143 (K; LISC; LMA). MS: Manica, Revuè near Macequece Rd., fl. & fr. 10.iii.1948, *Garcia* 556 (LISC; LMA; WAG). GI: Gaza, from Guijá towards Mabalane, near Aldeia da Barragem, fl. & fr. 3.vi.1969, *Barbosa & Lemos* 8579 (COI; K; LISC; LMA). M: Maputo, Bela Vista, between Tinonganine and Licuati Forest, fl. & fr. 8.xii.1961, *Lemos & Balsinhas* 266 (BM; COI; K; LISC; LMA).

Widespread throughout tropical and subtropical regions of both hemispheres. Woodlands, grasslands, savanna, edges of thicket, roadsides, sandy soils and as weed of cultivated ground; 0–1500 m.

4. SEDDERA Hochst.

Seddera Hochst. in Flora **27**, Beih: 7, t. 5 (1844).

Small prostrate to erect shrubs or suffrutices, neber climbing, intricately branched, more or less pubescent; the old branches sometimes becoming spinescent. Leaves variously shaped, entire, small, usually with medifixed hairs. Flowers 5-merous, axillary, solitary or aggregated into few-flowered clusters or pedunculate dense few-flowered cymes or in terminal spines or panicles. Bracteoles usually small. Sepals acute or obtuse, subequal or the outer ones slightly larger. Corolla small, funnel-shaped, the limb usually shallowly lobed; the midpetaline areas hairy at least at the tips. Stamens inserted low down in the corolla tube; filaments filiform, dilated at the base and often appendaged; anthers oblong; pollen smooth. Disk absent or small. Ovary 2-locular, 4-ovuled, hairy above; style bifid almost or quite to the base; stigmas more or less peltate and orbicular, sometimes bilobate. Capsule 4-valved. Seeds dark brown or black, glabrous.

A genus of about 20 species mainly restricted to Africa but extending to Arabia and India.

Flowers always solitary, sessible or very rarely pedicellate, calyx 7–10 mm. long, the sepals ciliate with bulbous-based hairs; corolla more or less 10 mm. long, pubescence usually distinctly brown or ferrugineous with bulbous-based hairs -   .   -   .   .   1. *capensis*
Flowers usually in few-flowered axillary cymes, rarely all solitary, sessile or pedunculate, calyx usually only 4–7 mm. long, corolla usually 6–7 mm. long; pubescence almost invariably white or grey without bulbous based hairs -   .   -   .   .   .   2. *suffruticosa*

1. **Seddera capensis** (E. Mey. ex Choisy) Hall. f. in Bull. Herb. Boiss. **6**: 529 (1898).—Baker & Wright in Dyer F.C. **4**: 80 (1904).—Baker & Rendle in F.T.A. **4**: 77 (1905).—Eyles in Trans. Roy. Soc. S. Afr. **5**: 452 (1916).—Meeuse in Bothalia **6**: 663 (1958).—Ross in Fl. Natal: 294 (1972). Type from S. Africa
  Evolvulus capensis E. Mey ex Choisy in DC., Prodr. **9**: 444 (1845).—Hall. f. in Engl., Bot. Jahrb. **18**: 86 (1893).
  Breweria capensis (E. Mey. ex Choisy) Baker & Wright in Dyer, F.C. **4**: 80 (1904). Type from S. Africa.
  Bonamia capensis (E. Mey. ex Choisy) Burtt Davy in Ann Transv. Mus. **3**: 121 (1912). Type from. S. Africa.

Suffruticose perennial. Stems several, tufted prostrate to suberect from a firm woody taproot, up to 30 cm. long, terete, covered with rusty brown appressed to patent stiff hairs as are the petioles, leaves, pedicels, calyces and midpetaline areas of the corolla. Leaves ovate, ovate-lanceolate or oblong, sessile or shortly petiolate, 5–22 × 3–12 mm., obtuse or subacute, minutely mucronate, rounded, to somewhat narrowed or truncate at the base, strigose on both surfaces, more laxly so when older, ciliate with bulbous-based hairs along the margin. Flowers axillary, solitary, usually subsessile, pedicels rarely up to 3 mm. long. Bracteoles 2, lanceolate, shorter than the sepals. Sepals broadly lanceolate, acute, 4–8 mm. long. Corolla broadly funnel-shaped, pinkish-white, 6–10 mm. long; mid-petaline areas with a few long strigose hairs. Ovary hairy at the top; style branched nearly from the base; stigmas subpeltate bilobed. Capsule subglobose, usually crowned with a tuft of hairs, about 5 mm. in diam. Seeds black, glabrous, smooth.

Botswana. SE: Mochudi, 3105 m., fl. & fr. v. 1914, *Rogers* 6850 (PRE). **Zimbabwe**. W: Bulawayo, fl. & fr. 15.iv.1965, *Gardner* 2 (K). C: Harare, fl. & fr. xii.1897, *Rand* 126 (BM). Also in S. Africa (Orange Free State, Natal and E. Cape Prov). Wet Veldt, up to 3105 m.

2. **Seddera suffruticosa** (Schinz) Hall. f. in Engl., Bot. Jahrb. **18**: 88 (1893) & in Bull. Herb. Boiss. **6**: 531 (1898).—Hiern, Cat. Afr. Pl. Welw. **1**: 725 (1898).—Baker & Rendle in F.T.A. **4**: 77 (1905).—Meeuse in Bothalia **6**: 663 (1958).—Verdc. in F.T.E.A., Convolvulaceae: 27 (1963).—Roessler in Merxm., Prodr. Fl. SW. Afr. **116**: 21 (1967).—Ross, Fl. Natal: 294 (1972).—Compton, Fl. Swaziland: 474 (1976). TAB. **4**. Type from S. Africa.
    Breweria suffruticosa Schinz in Verh. Bot. Ver. Prov. Brand. **30**: 275 (1888).–Baker & Wright in Dyer, F.C. **4**: 80 (1904). Type from S. Africa.

An extremely variable suffruticose or herbaceous perennial. Stems several to many from a woody rootstock, up to 60 cm., erect or spreading, terete or subterete, at first more or less densely covered with more or less whitish hairs to villous, glabrescent; branchelets appressed pubescent or with dense patent hairs. Leaves subsessile; leaf lamina from lanceolate to broadly elliptic-oblong, 7–45 × 3–13 mm. acuminate, acute or rounded at the apex and mucronulate, rounded at the base, pilose to densely hirsute on both surfaces with medifixed hairs; petiole up to 3 mm. long. Flowers in few-flowered hairy dense sessile or pedunculate axillary cymes, sometimes solitary; peduncles, up to 23 mm. long in our area, bearing one to several flowers; bracteoles shorter than the calyx, lanceolate. Sepals up to 8 mm. long, the three outer lanceolate, diversely acuminate, sometimes retrorse, inner ones slightly shorter, all with ovate more or less subcoriacepus bases and herbaceous acute apices, pubescent or hirsute. Corolla somewhat campanulate, white or yellowish, up to 6 mm. long, midpetaline area more or less densely pubescent. Ovary hairy at the apice. Capsule ovoid-subglobose, glabrous. Seeds globose, black, glabrous.

Plant not densely hairy   ·   ·   ·   ·   ·   ·   ·   ·   ·   ·   var. suffruticosa
Plant densely hairy especially the inflorescences, hirsute, somewhat more robust and with larger
    leaves   ·   ·   ·   ·   ·   ·   ·   ·   ·   ·   var. hirsutissima

Var. **suffruticosa**
    Convolvulus mucronatus Engl., Bot. Jahrb. **10**: 246 (1888). Type from. S. Africa.
    Seddera mucronata (Engl.) Hall. f. in Engl., Bot. Jahrb. **18**: 88 (1893). Type as above.
    Seddera welwitschii Hall. f. in Engl., Bot. Jahrb. **18**: 88 (1893); in Bull. Herb. Boiss.: **5**: 1009 (1897).—Hiern, Cat. Afr. Pl. Welw. **1**: 724 (1898).—Baker & Rendle in E.T.A. **4**: 77 (1905). Type from Angola.
    Breweria microcephala Baker in Kew Bull. **1894**: 68 (1894). Type from Angola.
    Breweria sessiliflora Baker in Kew Bull. **1894**: 68 (1894). Type from Mozambique: Zambezi Valley, between Sena and Lupata, *Kirk* (K, holotype).
    Breweria baccharoides Baker in Kew Bull. *1894*. 68 (1894). Type: Mozambique, Zambezi Valley, between Tete and the coast, *Kirk* (K, holotype).
    Seddera welwitschii var. *bakeri* Hiern, Cat. Afr. Pl. Welw. **1**: 725 (1898).—Baker & Rendle in F.T.A. **4**: 78 (1905). Type from Angola.
    Seddera welwitschii subsp. *tenuisepala* Verdc. In Kirkia **1**: 26, tab. II (1961). Type: Botswana, Sigara Pan, *Drummond & Seagrief* 5223 (K, holotype; LISC; SRGH, isotype).

Botswana. N: Gweta, main Rd., Maun-Francistown, iv.1967, *Lambrecht* 171 (K; SRGH). SW: Ghanzi, 945 m., fl. & fr. 12.i.1970, *Brown* 7753 (K). SE: Semowane R., NE. corner of Makarikari Pan, 910 m., fl. & fr. 23.iv.1957, *Drummond & Seagrief*, 5208 (K; SRGH). **Zambia**. S: Choma, fl. & fr. 15.v.1961, *Fanshawe* 6565 (K; LISC; SRGH). **Zimbabwe**. N: Binga, Kariba Research Station, fl. & fr. 25.ii.1966, *Jarman* 483 (K; SRGH). W: Matobo, Champion Ranch, c. 17 km. WNW. of Shashi-Shashani confluence, fl. & fr. 4.v.1963, *Drummond* 8083 (SRGH). E: Chipinge, E, Sabi, Giriwayo, fl. & fr. 19.i.1957, *Phipps* 26 (COI; K; SRGH). S: Mwenezi, Shirungwe Hill, 4.8 km. S. of Mateke Hills, 640 m., fl. & fr. 5.v.1958, *Drummond* 5586 (K; LISC; SRGH). **Malawi**. S: Nsanje Distr., between Thangadzi and Lilanje Rivers, fl. & fr. 25.iii.1960, *Phipps* 2688 (K; SRGH). **Mozambique**. T: Tete, from Tete to Changara, 200 m., fl. & fr. 21.iii.1966, *Torre & Correia* 15272 (BR; K; LMA). MS: Chemba, Chiou Experimental Station of CICA, fl. & fr. 9.iv.1962, *Balsinhas & Macuácua* 565 (K; LISC; LMA). GI: Gaza, Canicado, between Massingir and Kruger Park, fl. & fr. 13.xi.1970, *Correia* 1937 (LMU). M: Maputo, between Moamba and Ressano Garcia, fl. & fr. 3.xii.1940, *Torre* 2204 (LISC).
    Also in Tropical East Africa, Namibia and S. Africa. Open forest, mopane woodland, open grassy savanna, dry bush and scrub, stony and sandy soils and roadsides; 90–1370 m.

In agreement with Meeuse (tom. cit.) and Roessler (op. cit.) *S. welwitschii* has been sunk into *S. suffruticosa*. This decision is based on a large number of intermediate specimens which I have studied.

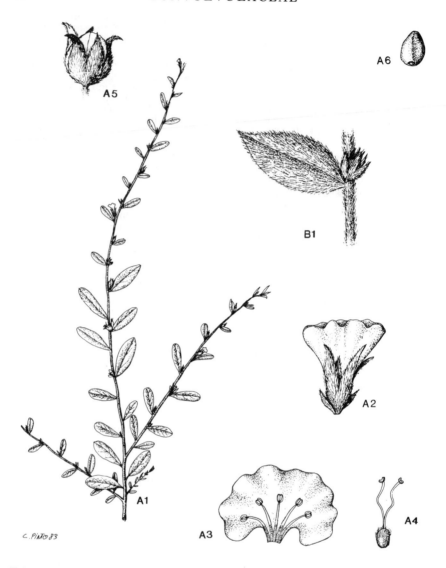

C. PINTO 83

Tab. 4. SEDDERA SUFFRUTICOSA var. SUFFRUTICOSA. A1, habit ($\times\frac{1}{2}$); A2, flower ($\times$3); A3, corolla opened to show stamens ($\times$3), A1–3 from *Lambrecht* 161; A4, pistil ($\times$3); A5, fruit ($\times$3); A6, seed ($\times$5) A4–6 from *Brown* 7755; B1, portion of branch of S. SUFFRUTICOSA var. HIRSUTISSIMA ($\times$3), from *Kerfoot* 8028.

Var. **hirsutissima** Hall. f. in Bull. Herb. Boiss. **6**: 531 (1898).—Hiern, Cat. Afr. Pl. Welw. **1**: 725 (1898).—Baker & Rendle in F.T.A, **4**: 77 (1905). Type from Angola.
   Breweria conglomerata Baker in Kew Bull. **1894**: 68 (1894). Type as above.
   Seddera conglomerata (Baker) Hall. f. in Bull. Herb. Boiss. **5**: 1008 (1897).
   Breweria suffruticosa var. *hirsutissima* (Hall. f.) Baker & Wright in Thiselton-Dyer, F.C. **4**: 81 (1904). Type from Angola.

  Botswana. **N**: 78 km. N. of Aha Hills, fl. & fr. 13.iii.1965, *Wild & Drummond* 7002 (K; LISC; SRGH). **SW**: Mabuasefubi Pan, 116 km. S. of Tshane, fl. & fr. 24.ii.1960, *Wild* 5143 (K; SRGH). **SE**: Illalambele-Mosu area, nr. Soa Pan, fl. & fr. 11.i.1974, *Ngoni* 299 (K; SRGH).

Zimbabwe W: Nyamandhlovu Experimental Station, fl. & fr. 17.v.1973, *Denny* 4591 (SRGH).
S: Gwanda, 550 m., fl. & fr. 14.xii.1956, *Davies* 2279 (SRGH). **Mozambique**. T: Tete-
Changara Rd., 200 m., fl. & fr. 14.ii.1968, *Torre & Correia* 17614 (C; LISC; LMA; MO; WAG).

Also in Angola and S. Africa. Savanna, grassland, scrub on Kalahari sand and serpentine soil;
200–930 m.

5. BONAMIA Thouars

Bonamia Thouars, Hist. Veg. Afr. **1**: 17–18, 32–33, t. 8 (1804) & Dict. Sc. Nat. **5**:
145 (1806) *nom. conserv.*—T. Myint & D. B. Ward in Phytologia **17**, 3: 123–239
(1968)

Breweria R. Br., Prodr.: 487 (1810).

Shrubby climbers, rarely erect subshrubs. Leaves herbaceous or occasionally subcoriaceous, entire, lanceolate, ovate or elliptic. Flowers in dense, usually many-
flowered, axillary or terminal cymes or panicles; Bracteoles usually small. Sepals 5,
equal or subequal, rarely very unequal, circular to lanceolate, herbaceous or coriaceous,
never membranous, not accrescent. Corolla funnel-shaped, medium or small sized,
blue or white, 5-lobed, with midpetaline bands hairy outside; lobes induplicate-
valvate. Stamens 5, included or slightly exserted; filaments oftend dilated and hairy
below or glabrous; anthers oblong, cordate or sagittate at the base; pollen smooth.
Ovary 2-locular, 4-ovuled, glabrous or hairy at apex; style bifid often with two unequal
branches or two styles nearly or quite free; stigmas 2 globose or peltate; disk small or
absent. Capsule subglobose, bilocular, 4-valved, 4–2-seeded.

A genus of about 40 species widely spread in the tropics.

A member of the allied genus *Porana*, namely *Porana paniculata* Roxb., "Horsetail liane",
native from India, a strong shrubby climber with panicles terminating every branchlet with
inumerable white flowers, is cultivated in Zambia, Zimbabwe and Mozambique.

1. Corolla blue, more than 1·5 cm. long; stigmas globose   -   -   -   -   -   -   2
— Corolla white, up to 1·5 cm. long; stigmas peltate   -   -   -   -   -   3. *velutina*
2. Leaves up to 11·5 cm. long; indumentum of dense long spreading hairs (drying golden-
brown)   -   -   -   -   -   -   -   -   -   -   -   1. *mossambicensis*
— Leaves up to 8.5 cm. long; indumentum short and appressed   -   -   -   2. *spectabilis*

1. **Bonamia mossambicensis** (Klotzsch) Hall. f. in Engl., Bot. Jahrb. **18**: 91 (1893).—
Dammer in Engl., Pflanzenw. Ost.-Afr. **C**: 328 (1895).—Baker & Rendle in F.T.A. **4**: 79,
fig. 8 (1905).—Brenan, T.T.C.L.: 169 (1949).—Verdc. in F.T.E.A., Convolvulaceae: 29,
fig. 8 (1963). Type: Mozambique, Sena, *Peters* (B, holotype†).

Prevostea mossambicensis Klotzsch in Peters, Reise Mossamb. Bot. **1**: 244, t. 39 (1861).
Type as above.

Breweria buddleoides Baker in Kew Bull. **1894**: 69 (1844). Type from Tanzania or
Mozambique: banks of R. Rovuma, 48 km. inland, *Kirk* (K, holotype).

Shrubby climber up to 5 m. or more. Stems woody, velvety with patent and tangled
hairs (white or grey in life but golden brown when dry). Leaves subcoriaceous, shortly
petioled; leaf-lamina elliptic-lanceolate to oblong-ovate, 2·5–11·5 × 1–4·3 cm., acute
and mucronulate or apiculate at the apex, rounded or slightly cordate at the base,
velvety pubescent above, densely coated with brown hairs beneath; petiole up to 1·0
cm. long. Flowers in bracteate densely hirsute, capitate inflorescences; peduncles up to
4 cm. long, axillary, many-flowered; bracts 1–1·5 × 0·5–0·8 cm., hirsute outside and
margin with a rusty tomentum, glabrous inside. Sepals unequal, up to 9 mm. long,
elliptic-oblong, acute, glabrous except the apex of the outer ones. Corolla bright blue,
funnel-shaped, up to 3 cm. long, not distinctly lobed, pilose outside. Ovary hairy at the
apex; style unequally bifid above the middle; stigmas ovoid. Capsule globose-oblong,
pilose. Seeds oblong, brownish, edged with narrow hyaline golden wings.

Mozambique. N: Niassa, between Espozende and Diaca, fl. & fr. 14.xi.1953, *Balsinhas* 111
(BM; LISC; LMA). Z: Maganja da Costa, Bajone, Murroa, fl. & fr. 20.v.1971, *Balsinhas* 1890
(LMA). MS: 36 km. from Maringa, Sabi R., fl. & fr. 18-vi.1950, *Chase* 2552 (BM; K; LISC;
SRGH). GI: Macovane, 7.2 km. N. of Inhassoro, fl. & fr. 10.x.1963, *Leach & Bayliss* 11900 (K;
LISC; SRGH).

Also in Tanzania. Open woodland, *Brachystegia* thicket, thick and coastal bush, savanna, edge
of forest and roadside; 10–800 m.

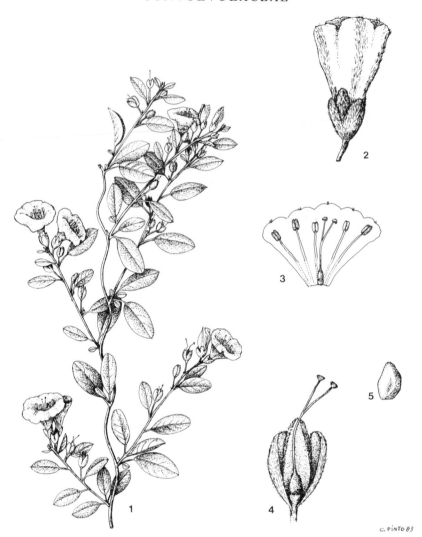

C. PINTO 83

Tab. 5. BONAMIA SPECTABILIS. 1, habit (×½); 2, flower (×2); 3, corolla opened to show stamens and pistil (×1); 4, fruit with two sepals removed (×3); 5, seed (×10). All from *Lawton* 208.

2. **Bonamia spectabilis** (Choisy) Hall. f. in Engl., Bot. Jahrb. **16**: 529 (1893); Engl., Bot. Jahrb. **18**: 91 (1893).—Verdc. in F.T.E.A., Convolvulaceae: 31 (1963). TAB. **5**. Type from Madagascar.

    Breweria spectabilis Choisy in Mem. Soc. Phys. Genève **8**: 68 (1839); DC., Prodr. **9**: 439 (1845). Type as above.

    Breweria hildebrandtii Vatke in Linnaea **43**: 523 (1882). Type from Madagascar.

    Bonamia hildebrandtii (Vatke) Hall. f. in Engl., Bot. Jahrb. **18**: 91 (1893). Type as above.

    Bonamia minor Hall. f. in Engl., Bot. Jahrb. **18**: 91 (1893);—De Wild. in Ann. Mus. Congo, Bot. Sér. 4: 110 (1903).—Baker & Rendle in F.T.A. **4**: 80 (1905). Type from Zaire.

    Bonamia minor var. *argentea* R. E. Fries, Wiss. Ergebn. Schwed. Rhod.-Kongo-Exped. **1**: 268 (1916). Type: Zambia, Lake Bangweulu, *Fries* 827 (UPS, holotype; K, isotype).

Climbing shrub with woody rootstock and numerous small shoots. Stems ridged, sparsely appressedly pubescent. Leaves petioled; lamina elliptic 2·5–8·5 × 0·8–3 cm., acute or obtuse and apiculate sometimes retuse, more or less acuneate at the base, glabrescent and green or silvery pubescent above, silky pubescent to glabrescent with glandular points beneath, fascicled in the leaf axils at the lower nodes; petiole up to 2·5 cm. long, slightly winged. Flowers in 1–3-flowered cymes aggregated towards the ends of the branchlets; peduncle and pedicels up to 16 mm. long; bracts small. Sepals oblong-elliptic 7–8 mm. long abruptly acute silky pubescent. Corolla bright blue, funnel-shaped, subentire, up to 23 mm. long, silky pilose on medpetaline areas. Ovary glabrous; style bifid from about the middle; stigmas ellipsoid. Capsule globose, shortly apiculate, glabrous. Seeds oblong in outline, blackish with narrow hyaline golden wings on the edges and with a reticulate pattern on the surface.

Zambia. B. Nangweshi, 1036 m., fl. & fr. 23.vii.1952, *Codd* 7156 (BM; K; PRE; SRGH). N: Mbala, Kituta Bay, 213 m., fl. & fr. 20.v.1936, *Burtt* 6325 (BM; K; SRGH). W: Kitwe, *Fanshawe* 7669 (K). C: 65 km. S. Lusaka, 914 m., fl. & fr. 24.v.1958, *Best* 131 (K; LISC; SRGH). S: Between Kafue R. and Kabwe, fl. & fr.13.vii.1930, *Hutchinson* 3599 (BM; COI; K; LISC; SRGH). **Zimbabwe**. N: Gokwe, 860 m., fl. & fr. 14.vi.1977, *Biegel* 5495A (SRGH). C: Golden Valley, fl. & fr. 8.iv.1933, *Michelmore* 688 (K).
 Also in Zaire, Tanzania and Madagascar. Mixed deciduous woodland and deciduous thicket, Kalahari Sand and roadsides; 825–1200 m.

3. **Bonamia velutina** Verdc. in Kirkia **1**: 27, tab. III (1961). Type: Zimbabwe, border opposite Chicualacuala, *Wild* 4688 (K, holotype; SRGH, isotype).

Perennial herb or suffrutex, erect, up to 0·9 m. high tomentose—Stems woody, branched, velvety with appressed sericeous hairs. Leaves shortly petiolate; lamina elliptic or elliptic-oblong, acute, 1–6·5 × 0·4–2·7 cm., mucronate at the apex, rounded or subtruncate at the base, greyish-velvety pubescent above, densely coated with golden brown hairs beneath; petiole up to 7 mm. long. Flowers in 1–3-flowered cymes; peduncles up to 5 mm. long; pedicles up to 2 mm. long; bracts leafy or minute. Sepals subequal, ovate-lanceolate, obovate or spathulate, subcoriaceous at the base, densely golden-velvety outside and the middle inside, glabrous at the base inside, pilose at the apex. Corolla white, funnel-shaped, slightly 5-lobed, up to 1·5 cm. long, silky pilose on midpetaline areas. Ovary ovoid, golden-pilose; style bifid below the middle; stigmas large lobulate-peltate. Capsule ellipsoid, subacute appressed pilose at the apex and base. Seeds ellipsoid, brown-purple, glabrous, minutely punctate.

Botswana. SE: Mochudi, 5.4 km. SE of Khomo, fl. & fr. x.1947, *Miller* 473 (FHD). **Zimbabwe**. S: Mwenezi, fl. & fr. 29.iv.1962, *Drummond* 7809 (K; SRGH). **Mozambique**. GI: Gaza, Massingir, fl. & fr. 24.vii.1982, *Cardoso de Matos* 5090 (LISC). M: Maputo, fl. & fr. vi.1914, *Maputoland Expedition* 14243 (PRE).
 Also in S. Africa (Natal). Open and mixed woodland, shrubby thicket, Kalahari Sand and sandstone; 200–980 m.

6. CRESSA L.

Cressa L., Sp. Pl.: 223 (1753); Gen. Pl., ed. 5: 104 (1754).

A much-branched, small, short-lived, perennial, subshrub-like herb. Leaves small, entire, sessile. Flowers small, subsessile, aggregaged in brateate clusters at the ends of the branchlets. Sepals 5, obovate, subequal, imbricate, coriaceous. Corolla 5-lobed; tube campanulate; lobes ovate, imbricate in bud, spreading. Stamens exserted; filaments filiform, glabrous, anthers oblong. Ovary 2-locular, 4-ovuled. Styles 2, exserted, distinct from the base; stigmas large, capitate. Capsule 2–4-valved, usually single-seeded. Seed glabrous and shining, dark brown.

Probably a monotypic genus, throughout the tropics.

Cressa cretica L., Sp. Pl.: 223 (1753).—Peter in Engl. & Prantl, Pflanzenfam. ed. 1, **4**, 3a: 17 (1891).—Hall. f. in Engl., Bot. Jahrb. **18**: 87 (1893).—Hiern, Cat. Afr. Pl. Welw. **1**: 724 (1898).—Baker & Rendle in F.T.A. **4**: 72 (1905).—Dandy in F. W. Andr., Fl. Pl. Anglo-Egypt. Sudan **3**: 108 (1956).—Heine in F.W.T.A. ed. 2, **2**: 339 (1963).—Verdc. in F.T.E.A., Convolvulaceae: 33 (1963).—Siddigi in Fl. Libya **45**: 2 (1977). TAB. **6**. Type from Crete.

Stems woody at base, terete, slender, with numerous spreading or ascending, hairy, densely-leaved branchlets. Leaves sessile, ovate-lanceolate to ovate, 2–7 × 1–4 mm.,

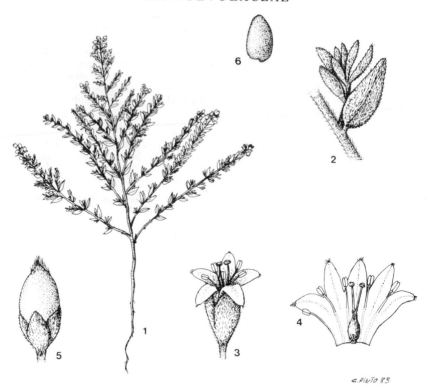

Tab. 6. CRESSA CRETICA. 1, habit (×½); 2, portion of branch (×3); 3, flower (×4); 4, corolla opened to show stamens and pistil (×4); 5, fruit (×4); 6, seed (×4). All from *Mendonça* 3245.

acute at the apex, cuneate, rounded or even subcordate at the base, grey-green appressedly pubescent. Flowers aggregated at the ends of the branchlets. Calyx ovoid 3–4 mm. long. Sepals concave, abruptly acute, silky pubescent. Corolla about 5 mm. long; tube and lobes about equal, the first enveloped by the calyx; lobes oblong-ovate, hairy on the outside. Stamens and styles rather longer than the corolla. Capsule ovoid, pilose at the apex. Seed ovoid.

Mozambique. N: Mocimboa do Rovuma, fl. & fr. 23.ix.1948, *Pedro & Pedrogão* 5329 (LMA). GI: Gaza, Banhine Plains, fr. xi. 1973, *Tinley* 3002 (K; LISC; SRGH).
Sudan, Somali Republic, Kenya, Socotra, Senegal and Angola. Throughout the tropics of both the Old and New World. In saline soil.

7. JACQUEMONTIA Choisy

Jacquemontia Choisy in Mem. Soc. Phys. Genève 6: 476 (1834).—Thyella Raf., Fl. Tell. 4: 84 (1838).

Herbaceous to woody twining or prostrate herbs or shrubs, rarely erect, mostly hairy, with 2–3-armed hairs. Leaves variable, entire often cordate at the base. Inflorescences usually bracteate. Flowers in axillary usually pedunculate cymes, or capitate or solitary; bracteoles small, linear to lanceolate or the outer ones larger, foliaceous; pedicels very short or flowers sessile. Sepals 5, equal or unequal with the 2 outer ones often large. Corolla small to medium-sized, funnel-shaped or campanulate, blue, mauve or pink, rarely white, obscurely 5-lobed; midpetaline areas distinct. Stamens included; filaments linear; anthers oblong; pollen smooth. Disk small or absent. Ovary bilocular, 4-ovuled; style simple, filiform, included; stigmas 2, elliptic or oblong, more or less flattened, rarely globose or filiform. Capsule globose, bilocular, 4-seeded or less by abortion. Seeds usually glabrous.

A large genus of about 120 species, mainly American, a few in tropical Africa, Asia and Australia.

1. Leaves and stems glabrous or nearly so - - - - - - 1. *ovalifolia*
— Leaves and stems distinctly pubescent or hairy - - - - - - 2
2. Flowers in few to many-flowered cymes; bracts minute - - - - 2. *paniculata*
— Flowers crowded into bracteate heads; bracts 2 cm. long - - - 3. *tamnifolia*

1. **Jacquemontia ovalifolia** (Vahl) Hall. f. in Engl., Bot. Jahrb. **18**: 96 (1893).—Hiern, Cat. Afr. Pl. Welw. **1**: 726 (1898).—Baker & Rendle in F.T.A. **4**: 87 (1905).—Heine in F.W.T.A. ed., 2, **2**: 340 (1963).—Verdc. in F.T.E.A., Convolvulaceae: 34 (1963). —Robertson in Ann. Missouri Bot. Gard. **61**: 502, fig. 1–8 (1974). Type probably from Trinidad.
   Convolvulus ovalifolius Vahl, Eclog. Amer. **2**: 16 (1798). Type as above.
   Convolvulus coeruleus Schumach. & Thonn., Beskr. Guin. Pl.: 101 (1827). Type from West Africa.
   Ipomoea ovalifolia (Vahl) Choisy in Mem. Soc. Phys. Genève **6**: 449 (1834); DC., Prodr. **9**: 357 (1845). Type as for *Convolvulus ovalifolius*.
   Jacquemontia coerulea (Schumacher) Choisy ex G. Don, Gen. Syst. **4**: 283 (1838). Type as for *Convolvulus coeruleus*.
   Ipomoea oleracea Welw., Apont. Phyto-Geogr. in Ann. Cons. Ultr. **1**: 589 (1859). Type from Angola.

Annual or perennial herb. Stems slender, woody at base, prostrate-ascending, radiating from a thick rootstock, glabrous or nearly so, with 2-armed hairs. Leaf lamina oblong to oblong-lanceolate or elliptic to subcircular, up to 5 × 3·5 cm., obtuse or emarginate at the apex, cuneate at the base, more or less glabrous, very slightly fleshy; petiole up to 3 cm. long. Inflorescences few-flowered axillary cymes; peduncle up to 3·5 cm. long, erect Bracts linear or obovate, up to 8 × 3mm. Sepals ovate or obovate-elliptic, acute, unequal, 4–7 × 4 mm. venose in fruit. Corolla blue, subrotate to shallowly campanulate, about 1 cm. long. Capsule pale brown, thin, subglobose; fruiting pedicels up to 1·3 cm. long. Seeds trigonous, minutely areolate, the outer 2 margins narrowly and irregularly winged.

Mozambique. G I: Gaza, Banhine National Park, fl. x.1973, *Tinley* 2990 (SRGH).
East and West Africa, Angola and Namibia; also in America. Floodplain depressions.

2. **Jacquemontia paniculata** (Burm. f.) Hall. f. in Engl., Bot. Jahrb. **16**: 541 (1893); Engl., Bot. Jahrb. **18**: 95 (1893).—Dammer in Engl., Pflanzenw. Ost.-Afr. **C**: 329 (1895).—Baker & Rendle in F.T.A. **4**: 87 (1905).—Verdc. in F.T.E.A., Convolvulaceae: 34 (1963).—Ross, Fl. Natal: 294 (1972). TAB. 7. Type from Java.

Var. **paniculata**.
   Ipomoea paniculata Burm. f., Fl. Ind.: 50, t. 21, fig. 3 (1768). Type as above.
   Convolvulus parviflorus Vahl. Symb. Bot. **3**: 29 (1794) non Desr. (1791) *nom. illegit*. Type from the East Indies.
   Convolvulus paniculatus (Burm. f.) O. Ktze., Rev. Gen. Pl.: 440 (1891), non L. (1753) *nom. illegit*.
   Jacquemontia parviflora (Vahl) Roberty in Candollea 14: 32 (1952) *nom. illegit*.

Annual twiner. Stems slender, wide-climbing, pubescent. Leaf lamina obovate-cordate, 3–4·5 ×2·5–3 cm., acuminate at the apex, truncate to subcordate at the base, glabrescent above and sparsely pubescent beneath on the nerves or densely tomentose all over with 3-armed hairs; petiole up to 42.5 cm. long, slender pubescent. Flowers several to many in a more or less dense cyme; peduncle up to 3 cm. long, slender, pubescent; secondary peduncles and pedicles up to 5 mm. long; bracts minute, deciduous. Sepals ovate, subequal, acute to acuminate, about 5 × 3 mm., pubescent. Corolla white, pink or yellow, more or less 10 mm. long. Capsule globose, straw-coloured, glabrous. Seeds subtrigonous, dark, with very narrow more or less indistinct hyaline margins and pale raised roughenings.

Tab. 7. JACQUEMONTIA PANICULATA. 1, habit (×½); 2, flower (×3); 3, corolla opened to show stamens (×3); 4, pistil (×3); 5, fruit (×3); 6, seed (×3). All from *Torre* 4100.

Mozambique. N: 16 km. W. of Lumbo, 1580 m., fl. & fr. 21.v.1961, *Leach & Rutherford-Smith* 10948 (K; SRGH). MS: Marromeu, between Lacerdónia and Chupanga, fl. & fr. 8.v.1942, *Torre* 4100 (C; LISC; LMA; MO; WAG).

Tropical East Africa, Madagascar to SE. Asia, Malaysia, tropical Australia and New Caledonia. In thickets and grassland; c. 1580 m.

The var. *tomentosa* (Warb.) Ooststr. and the var. *phillipinensis* Ooststr. are recorded from Malaysia.

3. **Jacquemontia tamnifolia** (L.) Griseb., Fl. Brit. W. Ind.: 474 (1861).—Peter in Engl. & Prantl, Pflanzenfam. ed. 1, **4**; 33(1891).—Dandy in F. W. Andr., Fl. Pl. Anglo-Egypt. Sudan **3**: 123 (1956).—Meeuse in Bothalia **6**: 700 (1958).—Heine in F.W.T.A. ed. 2, **2**: 340 (1963).—Verdc. in F.T.E.A., Convolvulaceae: 35, fig. 10 (1963).—Roessler in Merxm., Prodr. Fl. SW. Afr.**116**: 18 (1967).—Binns, H.C.L.M.: 40 (1968).—Munday & Forbes, First Check List Fl. Inhaca Is., Mocamb.: Based on col. A.O.D. Mogg in Journ. S. Afr. Bot. **45**: 9 (1979). Type from N. America.

Ipomoea tamnifolia L., Sp. Pl.: 162 (1753). Type as above.
Convolvulus capitatus Desr. in Lam., Encycl. **3**: 554 (1791). Type from Senegal.
Convolvulus guineensis Schumach in Schumach & Thonn., Besk. Guin. Pl.: 90 (1827). Type from Ghana.
Jacquemontia capitata (Desr.) G. Don, Gen. Syst. **4**: 283 (1837).—Peter in Engl. & Prantl, Pflanzenfam. ed. 1, **4**, 3a:33 (1891).—Dammer in Engl., Pflanzenw. Ost.-Afr. **C**: 329 (1895).—Hiern, Cat. Afr. Pl. Welw. **1**: 725 (1898).—Baker & Wright in Dyer, F.C. **4**: 69 (1904).—Baker & Rendle in F.T.A. **4**: 85 (1905).—Eyles in Trans. Roy. Soc. S. Afr. **5**: 452 (1916).—Hutch & Dalz., F.W.T.A. **2**: 211 (1931).—Wild, Guide Fl. Victoria Falls: 155 (1953).
Ipomoea guineensis (Schumach.) G. Don., Gen. Syst. **4**: 269 (1837). Type as for *Convolvulus guineensis*.
Thyella tamnifolia (L.) Raf., Fl. Tell. **4**: 84 (1838). Type as for *Ipomoea tamnifolia*.
Ipomoea capitata (Desr.) Choisy in DC., Prodr. **9**: 365 (1845). Type as for *Convolvulus capitatus*.
Convolvulus pycnanthus Choisy in DC., Prodr. **9**: 365 (1845). Type from the Sudan Republic.

Annual twiner. Stems several from the base, twining or trailing, occasionally suberect, up to about 85 cm. long, appresed pilose with silky hairs, usually brownish. Leaf lamina ovate, oblong to broadly cordate, up to 9 × 6 cm., acute to acuminate at the apex, truncate to cordate at the base, glabrescent or more or less pilose with brownish or white hairs, ciliate; petiole slender, up to 4·5 cm. long, usually much more densely pilose than the blade or stem. Flowers in dense hairy capitate cymes more or less 2·5 cm. across, of a ferrugineous colour when dry, supported by reduced bract-like leaves with narrowing base, the inner ones becoming smaller, lanceolate or linear, and more hairy ultimately resembling the sepals; peduncles up to 12 cm. long, silky pilose above. Sepals subequal, more or less 5 mm. long, lanceolate, acute, densely and softly hairy with ferrugineous or rarely white hairs. Corolla blue, more or less 1 cm. long, funnel-shaped, obscurely 5-lobed, glabrous, fugacious. Capsule globose, glabrous, straw-coloured. Seeds usually 4, brown scabridulous.

Botswana. N: Shakawe, fl. & fr. 24.iv.1975, *Biegel, Müller & Gibbs Russell* 4995 (K; SRGH). **Zambia**. B: Lusu, fl. & fr. 9.vii.1962, *Fanshawe* 6925 (SRGH). C: 25 km. W. of Lusaka on Mumbwa Rd., fl. & fr. 20.iii.1965, *Robinson* 6448 (K; SRGH). E: Luangwa R., 520 m., fl. & fr. 25.iii.1955, *Exell, Mendonça & Wild* 1189 (BM; LISC; SRGH). S: Livingstone, fl. & fr. 21.ii.1963, *Lawton* 1040 (K; SRGH). **Zimbabwe**. N: Gokwe, Ungwe R., fl. & fr. 20.iii.1964, *Bingham* 1282 (K; SRGH). W: Hwange, fl. & fr. 22.vi.1934, *Eyles* 8084 (BM; K; SRGH). C: 126 km. from Harare, *Craster* 1916 (K). E: Mutare, nr. Odzi R., 945 m., fl. & fr. 31.vi.1956, *Chase* 6137 (BM; COI; K; LISC; SRGH). S: Mwenezi, Malangwa R., SW. Meteke Hills, 625 m., fl. & fr. 6.v.1958, *Drummond* 5605 (K; LISC; SRGH). **Malawi**, N: Karonga, 472 m., fl. & fr. 20.iv. 1972, *Pawek* 5134 (K; SRGH). C: Salima, Senga Bay, 472 m., fl. & fr. 24.iv.1971, *Pawek* 4717 (K). S: Mulundunzi, fl. & fr. 15.vii.1952, *Williamson* 27 (BM). **Mozambique**. N: Nampula, Experimental Station CICA, fl. & fr. 7.iv.1961, *Balsinhas & Merrime* 351 (BM; COI; K; LISC; LMA; SRGH). Z: Namagoa, Mocuba, 70–274 m., fl. & fr. iv.1945, *Faulkner* 202 (K; SRGH). T: Tete, Changara Rd., 200 m., fl. & fr. 19.iii.1966, *Torre & Correia* 15244 (BR; K; LISC; M). MS: Bandula, 700 m., fl. & fr. 5.iv.1952, *Chase* 4450 (BM; K., SRGH). GI: Between Nhachengo and Vilanculos, fl. & fr. 22.iii.1952, *Barbosa & Balsinhas* 4975 (BM). M: Maputo, Marracuene, Vila Luisa, fl. & fr. 29.vi.1961, *Balsinhas* 496 (BM; COI; K; LISC; LMA; SRGH).

Tropical and S. Africa, Madagascar, the Mascarene Is. and tropical America. Woodland, edge of thicket, grassland, riverine woods, damp sandy ground, cultivated land; 60–1100 m.

8. CONVOLVULUS L.

Convolvulus L., Sp. Pl.: 153 (1753); Gen. Pl., ed. 5: 76 (1754).

Annual or perennial herbs with erect, prostrate or twining stems. Leaves petiolate or nearly sessile, very variable, entire or lobed, often cordate, but often nastate or sagittate. Flowers axillary in 1-few-flowered cymes, sometimes subumbellate. Sepals 5, equal or subequal, persistent, obtuse to acute. Corolla funnel-shaped, white, pink or pinkish-purple, shallowly 5-lobed, midpetaline areas often hairy towards the apices. Stamens 5, usually unequal, included, filaments filiform or somewhat linear, often dilated at base; pollen smooth, ellipsoid. Ovary bilocular, 4-ovuled. Style simple, filiform, included; stigmas 2, filiform. Capsule bilocular, usually 4-valved. Seeds usually 4, black or brown, usually glabrous.

A large genus of about 250 species in the temperate and subtropical regions of both hemispheres with only a few in the tropics.

1. Corolla about 20 mm. long; introduced twining weed in cultivations   -   - 4. *arvensis*
—   Corolla 8–15 mm. long; indigenous plants -   ·   ·   ·   ·   ·   2
2. Leaf-lamina linear to linear-oblong, rhomboidal, ovate-oblong or palmately 5-fid; stems erect to decumbent, the suffruticose tomentose, more rarely pubescent   -   1. *ocellatus*
—   Leaf-lamina triangular-ovata, oblong or ovate, rarely lanceolate or linear; stems prostrate or twining, hairy, pubescent or farinose-puberulous ·   ·   ·   ·   ·   3
3. Leaf-lamina very variable, oblong, triangular-ovate to linear, up to 7 × 3 cm., sagittate at base with the lobes often bifid, entire, crenate or deeply laciniate; stems hairy   2. *sagittata*
—   Leaf-lamina triangular-ovate, rarely lanceolate, up to 11 × 16 cm., subsagittate or cordate at base, subentire to irregularly and shallowly crenate; stems pubescent or farinose-puberulous   ·   ·   ·   ·   ·   ·   ·   3. *farinosus*

1. **Convolvulus ocellatus** Hook. f., Bot. Mag. **70**: t. 4065 (1844).—Choisy in DC., Prodr. **9**: 404 (1845).—Hall. f. in Engl., Bot. Jahrb **18**: 102 (1894).—Baker & Wright in Dyer, F.C. **4**: 71 (1904).—Meeuse in Bothalia **6**: 672 (1958).—Roessler in Merxm., Prodr. Fl. SW. Afr. **116**: 4 (1967). Type (topotype) from S. Africa (Transvaal).

Erect to decumbent perennial forming much branched stems from a woody root-stock. Stems several from the base, densely brownish, greyish or sericeo-tomentose, more rarely white pubescent, up to about 60 cm. high. Leaf lamina linear to linear-oblong, rhomboidal, ovate-oblong or palmately 5-fid, sometimes emarginate or sagit-tate with basal auricles, often bifid, subtruncate or cuneate, usually acute, 9–35 × 1–25 mm., nearly sessile, thick, with stout midrib and the lateral nerves impressed above and prominent below, covered (as are peduncles, bracts, pedicels, calyx and midpetaline zones) with same brownish or greyish tomentum as the stems (often bullate or plicate in var. *ornatus*). Peduncles up to 5 mm. long, 1 (2) flowered, shorter than the leaves and up to about 30 mm. long; bracteoles small, linear or subulate, about 3 mm. long; pediclels usually longer than the peduncles and up to about 15 mm. long. Sepals about 6 (10) mm. long, oblong, elliptic or ovate, acute or somewhat acuminate, rarely obtuse, the outer ones completely covered with the tomentum on the outside, inner ones with a median hairy zone. Corolla white or pink, with a dark reddish-purple centre, widely funnel-shaped, about 12 mm. long. Capsule ovoid-conical, shortly apiculate, hairy at the apex. Seeds glabrous.

1. Leaves palmately 5-fid   ·   ·   ·   ·   ·   ·   ·   ·   ·   var. *ornatus*
—   Leaves undivided or at most coarsely serrate ·   ·   ·   ·   ·   ·   ·   2
2. Leaves emarginate or subsagittate with basal auricles   ·   ·   ·   ·   ·   3
—   Leaves without auricles, subtruncate or cuneate   ·   ·   ·   ·   ·   ·   4
3. Leaves linear to linear-oblong   ·   ·   ·   ·   ·   ·   ·   var. *ornatus*
—   Leaves rhomboidal or ovate-oblong   ·   ·   ·   ·   ·   ·   var. *plicinervius*
4. Leaves linear to linear-oblong   ·   ·   ·   ·   ·   ·   ·   var. *ocellatus*
—   Leaves linear-oblong only at base of stem, ovate-elliptic or rhomboidal above   ·   ·   ·   ·   ·   ·   ·   ·   ·   ·   var. *plicinervius*

Var. **ocellatus**.

Convolvulus randii Rendle in Journ. Bot. **40**: 189 (1902).—Baker & Rendle in F.T.A. **4**: 94 (1905).—Eyles in Trans. Roy. Soc. S. Afr. **5**: 452 (1916). Type: Zimbabwe, Gweru, *Rand* 274 (BM, holotype).

Zimbabwe. C: Gweru, Lalapanzi, fl. & fr. 22.i.1948, *Ingle* (SRGH).
Also in S. Africa (Transvaal). In woods and on red sandy soil; 1400 m.

Var. **ornatus** (Engl.) Meeuse in Bothalia **6**: 673 (1958). Type from S. Africa.
    Convolvulus ornatus Engl. in Bot. Jahrb. **10**: 274 (1888).—Baker & Wright in Dyer F.C. **4**: 76 (1904). Type from S. Africa (Cape Province).
    Convolvulus multifidus Hall. f. in Engl., Bot. Jahrb. **18**: 102 (1893) non Thunb. (1794). Type from S. Africa.
    Convolvulus dinteri Pilger in Engl., Bot. Jahrb. **45**: 219 (1910).—Dinter in Feddes Repert. **16**: 240 (1919). Type from Namibia.

Zimbabwe. E: Mutare, Nymaganu Peak, Commonage, 1220 m., fl. & fr. 3.i.1960, *Chase* 7247 (BM; K; LISC; SRGH).
Also known from Namibia, S. Africa (Transvaal, Cape Province and the Orange Free State). Grassland; 1220 m.

Var. **plicinervius** Verdc. in Kirkia **1**: 28, tab. IV (1961).—Wild in Kirkia **5**: 78 (1965). Type: Zimbabwe, Mazoe Distr., *Wild* 3926 (K, holotype; EA; SRGH, isotypes).

Zimbabwe. N: Lomagundi, nr. Rodcamp Mine, fl. & fr. 20.x.1960, *Rutherford-Smith* 315 (K; LISC; SRGH). C: Ngezi, Battlefields Rd., on Great Dyke, fl. & fr. 15.i.1962, *Wild* 5586 (K; SRGH). S: Shabani, fl. & fr. 17.iii.1972, *Wild* 7916 (K; LISC; SRGH).
Not known elsewhere. Open grassland, on serpentine soil; 1220–1615 m.

2. **Convolvulus sagittatus** Thunb., Prodr. Fl. Cap.: 35 (1794).—Choisy in DC., Prodr. **9**: 407 (1845).—Hall. f. in Engl., Bot. Jahrb. **18**: 103 (1893).—Dammer in Engl., Pflanzenw. Ost.-Afr. C: 329 (1895).—Hall. f. Bull. Herb. Boiss. **6**: 533 (1898).—Hiern, Cat. Afr. Pl. Welw. **1**: 726 (1898).—Hall. f. in Warb., Kunene-Samb.-Exped. Baum.: 345 (1903).—Baker & Wright in Dyer F.C. **4**: 72 (1904).—Baker & Rendle in F.T.A. **4**: 96 (1905), pro parte excl. var. *abyssinicus*.-Rendle in Soc., Bot. **40**: 150 (1911).-Eyles in Trans. Roy. Soc. S. Afr. **5**: 452 (1916).—Fries, Wiss. Ergebn. Schwed.-Rhod.-Kongo-Exped. **1**: 268 (1916).—Meeuse in Bothalia **6**: 679 (1958).—Verdc. in F.T.E.A., Convolvulaceae: 43 (1963).—Roessler in Merxm., Prodr. Fl. SW. Afr. **116**: 5 (1967).—Binns, H.C.L.M.: 39 (1968).—Ross, Fl. Natal: 295 (1972).—Jacobsen in Kirkia **9**: 171 (1973). Type from S. Africa.

An extremely variable perennial. Stems prostrate or twining, hairy, radiating for up to 0·6 m. from a woody rootstock. Leaf lamina very variable, oblong, triangular-ovate to linear 2·5–7 × 0·6–3 cm., acute at the apex, sagittate at the base, with lobes often bifid, entire, obscurely crenate or deeply laciniate, pubescent or glabrescent, often villous when young, petiole up to 15 mm. long, usually very short and often hairy. Inflorescence 1–3-(several) flowered; peduncle up to 4.5 cm. long, usually terete, slender, often hairy; pedicels about 5 mm. long; bracts small, linear. Sepals varying from lanceolate to circular, ovate or elliptic, up to 9 mm. long, acute to obtuse, sometimes mucronate or ciliate, the outer longer or narrower particularly in fruit, or sepals subequal, hairy to glabrous. Corolla white or sometimes with pink or purple-red centre or pink with a purple centre, 8–15 mm. long, midpetaline areas hairy near the apices. Capsule subglobose, glabrous. Seeds usually 4, dark brown or black, scabridulous more or less glabrous when ripe.

1. Leaves narrowly oblong to ovate, auriculate, almost truncate or hastate to sagittate, often with conspicuous bifid lobes at the base, irregularly undulate to laciniate; sepals ovate-lanceolate, subcircular-ovate, acute to acuminate, sericeous, glabrous; plants with young parts more or less golden-villous, older parts glabrescent; stems prostrate or twining
                                                   var. *aschersonii*
— Leaves not as above · · · · · · · · · · · ·   2
2. Leaves linear-sagittate with entire, rounded or rarely 2-lobed basal auricles; sepals ovate or broadly ovate, acute, hairy to nearly glabrous; plants thinly hairy to nearly glabrous with appressed hairs; stems prostrate, rarely climbing · · · ·   var. *sagittatus*
— Leaves sagittate or oblong-sagittate or somewhat hastate; basal auricles entire; sepals subspathulate, elliptic or obovate, obtuse, mucronate, with crisped margin, usually quite glabrous; plants usually densely and shortly pubescent on stems and petioles; stems prostrate · · · · · · · · ·   var. *phyllosepalus*

Var. **sagittatus**
    Convolvulus ulosepalus Hall. f. in Engl., Bot. Jahrb **18**: 103 (1893).—Baker & Wright in Dyer, F.C. **4**: 73 (1904).—Baker & Rendle in F.T.A. **4**: 95 (1905).—Engl. in Sitz.-Ber. Königl. Preuss. Akad. Wiss. Berl. **1907**: 26 (1907).—Eyles in Trans Soc. S. Afr. **5**: 452 (1916).—Meeuse in Bothalia **6**: 678 (1958).—Ross, Fl. Natal: 295 (1972).—Compton, Fl. Swaziland: 475 (1976). Type from S. Africa (Cape Prov.).
    Convolvulus thomsonii Baker in Kew Bull. **1894**: 67 (1894). Type from Tanzania.

Convolvulus sagittatus Thunb. subvar. *villosus* Hall. f. in Bull. Herb. Boiss. **6**: 533 (1898). Type as for *C. thomsonii*.

Convolvulus sagittatus subvar. *australis* Hall. f. in Bull. Herb. Boiss. **6**: 533 (1898). Type from S. Africa.

Convolvulus sagittatus Thunb. subvar. *linearifolia* Hall. f. in Bull. Herb. Boiss. **6**: 534 (1898); in Warb., Kunene-Samb.-Exped. Baum, Bot. Ser. **4**: 345 (1903). Type from S. Africa (Cape Prov.).

Convolvulus sagittatus Thunb. var. *linearifolius* (Hall. f.) Baker & Wright in Dyer, F.C. **4**: 72 (1904).—Baker & Rendle in F.T.A. **4**: 97 (1905).—Meeuse in Bothalia **6**: 683 (1958). Type from S. Africa.

Convolvulus sagittatus Thunb. var. *villosus* (Hall. f.) Rendle in F.T.A. **4**: 96 (1905). —Verdc. in F.T.E.A., Convolvulaceae: 43 (1963). Type as for *C. thomsonii*.

Convolvulus sagittatus Thunb. var. *ulosepalus* (Hall. f.) Verdc. in Kew Bull. **12**: 346 (1957); in F.T.E.A., Convolvulaceae: 44 (1963). Type from S. Africa.

Botswana. N: Ngamiland, fl. & fr. xii.1930, *Curson* 504 (PRE). SW: Ghanzi, farm 56, 950 m., fl. & fr. 19.i.1970, *Brown* 7952 (K; SRGH). SE: Molepolole, Honyi Pan, fl. & fr. 17.ii.1960, *Yalala* 33 (K; LISC; SRGH). **Zambia** B: Masese, fl. & fr. 12.i.1961, *Fanshawe* 6119 (SRGH). W: Kitwe, fl. & fr. 10.xi.1968, *Fanshawe* 10418 (SRGH). C: Lusaka, 1220 m., fl. & fr. 12.vii.1955, *Best* 107 (K). S: Mumbwa, Naleza, fl. & fr. 24.vi.1963, *van Rensberg* 2299 (K; LISC; SRGH). *Zimbabwe.* N: Mazoe, fl. & fr. 11.iii.1968, *Heberden* in GHS 187587 (K; LISC; SRGH). W: Hwange National Park, fl. & fr. 14.xi.1968, *Rushworth* 1264 (K; SRGH). C: Harare, between Avondale West and Mabelreign, fl. & fr. 23.x.1955, *Drummond* 4925 (K; SRGH). E: Inyanga, 1830 m., fr. 21.x.1935, *Eyles* 8473 (K; SRGH). **Malawi.** N: Mzimba, Kasito R., 32 km. W. of Mzuzu, 1160 m., fl. & fr. 8.vii.1974, *Pawek* 8799 (K; SRGH). C: Lilongwe 1050 m., fl. & fr. 26.vi.1970, *Brummitt* 11696 (K; LISC; MAL; PRE; SRGH; UPS). **Mozambique**. T: Angonia, Ulongue, fl. & fr. 29.xi.1980, *Macuácua* 1333 (LMA).

Also Sudan to West Africa and S. Africa. Open woodland, grassland, growing in pan and roadside; 940–2260 m.

Var. **aschersonii** (Engl.) Verdc. in Kew Bull. **12**: 345 (1957); in F.T.E.A., Convolvulaceae: 43 (1963).—Binns, H.C.L.M.: 39 (1968).—Jacobsen in Kirkia **9**: 171 (1973). Type from Ethiopia.

Convolvulus aschersonii Engl., Hochgeb. Trop. Afr.: 349 (1891).—Meeuse in Bothalia **6**: 677 (1958).—Heine in F.W.T.A. ed. 2, **2**: 340 (1963). Type from Ethiopia.

Convolvulus sagittatus Thunb. subvar. *abyssinicus* Hall. f. in Bull. Herb. Boiss. **6**: 533 (1898), excl. syn. *C. penicellatus* A. Rich. Types from, Ethiopia, Tanzania and Namibia (syntypes).

Convolvulus sagittatus Thunb. var. *abyssinicus* (Hall. f.) Baker & Rendle in F.T.A. **4**: 96 (1905). Eyles in Trans. Roy. Soc. S. Afr. **5**: 452 (1916).—Hutch & Dalz., F.W.T.A. **2**: 210 (1931). Type as above.

Convolvulus hallierianus Schulze-Menz in Notizbl. Bot. Gart. Berl. **14**: 377 (1939). —Brenan, T.T.C.L.: 169 (1949). Type from Tanzania.

Botswana. SW: 24 km. WNW. Hukuntsi along the track to Ncojane, fl. & fr. 28.x.1977, *Skarpe* S-201 (K; SRGH). **Zambia.** N: Mbala, Fwambo area on track from Kawimbe to Mbala, 1500 m., fl. & fr. 3.ix.1956, *Richards* 6071 (K). S: Choma, fl. & fr. 15.v.1961, *Fanshawe* 6566 (K; SRGH). **Zimbabwe.** N: Lomagundi, Mangula Township, 1180 m., fl. & fr. 16.ix.1962, *Jacobsen* 1772 (K; PRE). C: Gweru, 1400 m., fl. & fr. x.1919, *Eyles* 1820 (K; PRE; SRGH). **E**: Inyanga, 1700 m., fl. & fr. 30.x.1930, *Fries, Norlindh & Weimarck* 2442 (K; PRE; SRGH). **Malawi. N:** Vipya Plateau, 59 km. SW. of Mzuzu, 1680 m., fl. & fr. 12.vii.1975, *Pawek* 9853 (K; SRGH). **C**: Dedza, Kanjoli, Chongoni, fl. & fr. 4.xi.1967, *Salubeni* 863 (K; SRGH).

Widely distributed from Eritrea to S. Africa (Transvaal). Open woodland, grassland and savanna, roadside and very dry sand; 400–1800 m.

Var. **phyllosepalus** Hall. f. in Bull. Herb. Boiss. **6**: 535 (1898).—Baker & Wright in Dyer, F.C. **4**: 75 (1904). Type from S. Africa.

Convolvulus sagittatus Thunb. var. *latifolius* Wright in Dyer, F.C. **4**: 72 (1904). Type from S. Africa.

Zimbabwe. E: Inyanga, 2000 m., fl. & fr. 29.1.1931, *Norlindh & Weimarck* 4683 (K; PRE; SRGH). **Mozambique.** MS: Manica, Zuira Mt., Tsetserra, 2100 m., fl. & fr. 5.xi.1965, *Torre & Pereira* 12691 (LISC; LMA; WAG).

Also in S. Africa (Orange Free State, Natal and Transvaal). Montane grassland; 2000–2440 m.

3. **Convolvulus farinosus** L., Mant. **2**: 203 (1771).—Choisy in DC., Prodr. **9**: 412 (1845). —Peter in Engl. & Prantl, Pflanzenfam. **4**, 3a: 35 (1891).—Hall. f. in Engl., Bot. Jahrb. **18**: 104 (1893).—Baker & Wright in Dyer, F.C. **4**: 74 (1904).—Baker & Rendle in F.T.A. **4**: 98 (1905).—Meeuse in Bothalia **6**: 684 (1958).—Verdc. in F.T.E.A., Convolvulaceae: 41 (1963.—Compton, Fl. Swaziland: 474 (1976). TAB. **8**. Type from Sweden.

Convolvulus quinqueflorus Vahl, Symb. **3**: 31 (1794). Type from Réunion.

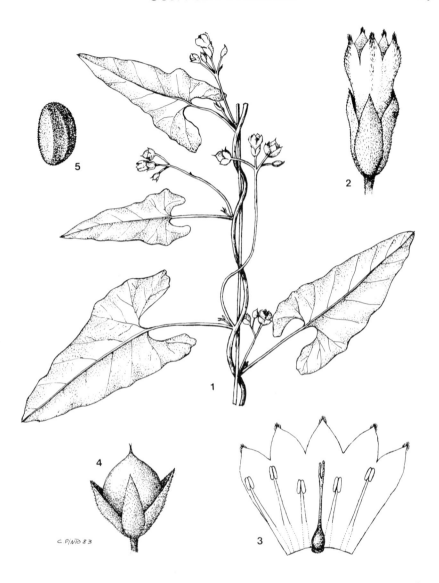

Tab. 8. CONVOLVULUS FARINOSUS. 1, habit ($\times\frac{1}{2}$); 2, flower ($\times3$); 3, corolla opened to show stamens and pistil ($\times3$); 4, fruit ($\times3$); 5, seed ($\times4$). All from *Barbosa & Lemos* 7861.

Convolvulus cordifolius Thunb., Prodr. Fl. Cap.: 35 (1794).—Choisy in DC., Prodr. **9**: 413 (1845). Type from S. Africa.
Convolvulus penicellatus A. Rich., Tent. Fl. Abyss. **2**: 74 (1851). Type from Ethiopia.
Convolvulus schweinfurthii Engl., Hochgeb. Trop. Afr.: 348 (1892). Type from Ethiopia.

Herbaceous perennial. Stems many, long and slender, twining or prostrate, pubescent or farinose-puberulous, the younger shoots often silvery. Leaf lamina triangular-ovate or ovate, rarely lanceolate, 3–11 × 3·8–6 cm., obtuse, acute or apiculate at the apex, subsagittate or cordate at the base, subentire to irregularly and shallowly crenate, herbaceous, drying membranous, glabrous above except when young, glabrous or more or less pubescent beneath and with distinct reticulate venation; basal sinus broad, basal auricles rounded or pointed, sometimes with a few teeth (very rarely leaves with additional triangular lobes above the basal auricles); petiole up to 6·5 cm. long. Inflorescence subumbellate, 1–6 flowered; peduncle up to 6 cm. long, finely pubescent; pedicels up to 1·5 cm. long; bracts minute. Sepals unequal, ovate-circular, ovate or elliptic, 5–9 × 3–5 mm., acute, more or less coriaceous, outer often pubescent all over; inner pubescent in the middle only, or all sepals ciliate, often purplish at the apices. Corolla white, very pale pink or pinkish-purple, 10–15 mm. long, shortly lobed, pubescent at the apices and the midpetaline areas, with the tube rather narrow, the lobes shortly mucronate at the apices. Capsule subglobose, shortly apiculate, glabrous. Seeds usually 4, subtrigonous, black, scabridulous-rugose.

Botswana. N: Ngamiland, fl. & fr. xii.1930, *Curson* 6 (PRE). **Zimbabwe**. C: Wedza Mt., fl. & fr. 7.iv.1964, *Wild* 6518 (K; LISC; SRGH). E: Mutare, Roonje Peak fl. & fr. 14.ix.1934, *Gilliland* 900 (BM; K; PRE; SRGH). S: Mberengwa, Buhwa Mt., 1400 m., fl. & fr. 31.x.1973, *Biegel, Pope & Gosden* 4342 (K; SRGH). **Malawi**. N: Nkhata Bay, Viphya Plateau, 1680 m., fl. & fr. 9.x.1976, *Pawek* 11875 (SRGH). C: Dedza Mt., fl. & fr. 1.xii.1967, *Salubeni* 914 (K; LISC; SRGH). S: Ntcheu, Chirobwe Hill, fl. & fr. 10.ii.1967, *Jeke* 69 (K; SRGH). **Mozambique**. GI: Chibuto, Xavanhane, fl. & fr. 16.viii.1957, *Barbosa & Lemos* 7861 (COI; K; LISC). M: From Bela Vista to Catuane, fl. & fr. 13.viii.1948, *Gomes e Sousa* 3785 (COI; K).
Tropical and subtropical eastern Africa, S. Africa, Madagascar, Mascarene Islands; also western Mediterranean area; 1000–2050 m.

4. **Convolvulus arvensis** L., Sp. Pl.: 153 (1753).—Choisy in DC., Prodr. **9**: 406 (1845). —Peter in Engl. & Prantl, Pflanzenfam. **4**, 3a: 35 (1891).—Hall. f. in Engl., Bot. Jahrb. **18**: 108 (1893).—Baker & Wright in Dyer F.C. **4**: 75 (1904).—Baker & Rendle in F.T.A. **4**: 97 (1905).—Dandy in F. W. Andr., Fl. Pl. Anglo-Egypt. Sudan **3**: 107 (1956).—Meeuse in Bothalia **6**: 695 (1958).—Siddigi in Fl. Lybia **45**: 6, fig. 2 (1977). Type from Sweden.

Perennial herb forming several to many annual stems from a long taproot. Stems prostrate or twining, angular, sparsely pubescent to glabrous. Leaf lamina entire, ovate-oblong or oblong, 1·5–4·5 × 0·5–1·5 cm., with a hastate or sagittate base, obtuse or mucronulate at the apex, glabrous or thinly hairy; petiole 5–10 mm. long. Flowers axillary, solitary or sometimes 2–3 flowered cymes; peduncles 1–1·5 cm.; bracts linear, about 2 mm. long; pedicels up to 10 mm. long. Sepals slightly unequal, elliptic-circular or oblong, 3.5–5 mm. long. Corolla white or pink or both, broadly funnel-shaped, about 2 cm. long in the limb shallowly lobed, glabrous except at apices of the midpetaline areas. Capsule ovoid-globose, glabrous, 4-seeded. Seeds dark brown or black.

Mozambique. M: Maputo, between Quinta da Pedra and Maputo R., near Salamanga, fl. & fr. 15.vii.1948, *Gomes e Sousa* 3752 (COI; K; LMA).
Originally a native of Europe and parts of Asia, now a common and widely distributed weed in the temperate and subtropical regions of both hemispheres, rarely in the tropics. Weed in cultivated soils.

9. HEWITTIA Wight & Arn.

Hewittia Wight & Arn. in Madras Journ. Sci. **5**: 22 (1837).
Shutereia Choisy in Mem. Soc. Phys. Genève **6**: 485 (1834) non *Shuteria* Wight & Arn. (1834) *nom. conserv.*
Eremosperma Chiov. in Fl. Somal. **3**: 143 (1936).—Verdc. in Webbia **13**: 321 (1958).

Twining or prostrate perennial herb. Flowers on axillary peduncles in 1-several-flowered bracteate cymes; bracteoles oblong or linear-lanceolate, acuminate, positioned at some distance from the calyx. Sepals 5, herbaceous, the outer ones large, accrescent in fruit, inner ones smaller. Corolla medium-sized, campanulate to funnel-shaped, heptangular. Stamens 5, included; filaments linear with dilated base; pollen smooth. Disk annular. Ovary hairy, unilocular or imperfectly bilocular at the apex, 4-ovuled; style simple, included; stigma 2, ovate-oblong, campanulate. Capsule locular, usually 4-valved, 4- or, by abortion, less-seeded. Seeds black, glabrous, opaque.

A small genus very close to *Convolvulus*, possibly with a single species widely spread in the tropics of the Old World and introduced into Jamaica and probably other areas of the New World.

Hewittia scandens (Milne) Mabberley in Manilal, Botany & History of Hort. Malab.: 84 (1980). TAB. **9** Type from India.
    Convolvulus sublobatus L.f., Suppl.: 135 (1781). Type from India.
    Convolvulus bicolor Vahl, Symb. **3**: 25 (1794) non Desr. (1791). *nom. illegit.* Type from Asia.
    Convolvulus involucratus Willd., Sp. Pl. **1**: 845 (1798). Type from Guinea.
    Shutereia bicolor (Vahl) Choisy in Mém. Soc. Phys. Genève **6**: 486, t. 2, f. 11 (1834). —Hiern, Cat. Afr. Pl. Welw. **1**: 727 (1898). Type from Asia.
    Hewittia bicolor (Vahl) Wight & Arn. in Madr. Journ. Sci. **5**: 22 (1837).—Peters, Reise Mossamb., Bot. **1**: 242 (1861.—Baker & Wright in Dyer, F.C. **4**: 68 (1904).—Baker & Rendle in F.T.A. **4**: 100 (1905).—Rendle in Journ. Linn. Soc., Bot. **40**: 150 (1911).—Eyles in Trans. Roy. Soc. S. Afr. **5**: 453 (1916). Type as above.
    Aniseia afzelii G. Don. Gen. Syst. **4**: 295 (1837).—Type from Sierra Leone.
    Hewittia asperifolia Klotzsch in Peters, Reise Mossamb., Bot.: 242 (1861). Type: Mozambique, Cabaceira Peninsula, near Mozambique, *Peters* (B, holotype†).
    Hewittia hirta Klotzsch in Peters, Reise Mossamb., Bot.: 243 (1861). Type: Mozambique, Sena, *Peters* (B, holotype†).
    Hewittia barbeyana Chodat & Roulet in Bull. Herb. Boiss. **1**: 191 (1893). Type from Senegambia.
    Hewittia sublobata (L. f.) Kuntze, Rev. Gen. Pl. **2**: 441 (1891).—Hall. f. in Engl., Bot. Jahrb. **18**: 111 (1893).—Dammer in Engl., Pflanzenw. Ost-Afr. **2**: 330 (1895). Brenan et al. in Mem. N.Y. Bot. Gard. **9**: 9 (1954).—Dandy in F. W. Andr., Fl. Pl. Anglo-Egypt. Sudan **3**: 109 (1956).—Meeuse in Bothalia **6**: 698 (1958).—Macnae & Kalk, Nat. Hist. Inhaca I., Mocamb.: 152 (1958).—Verdc. in F.T.E.A., Convolvulaceae: 45, fig. 12 (1963).—Binns, H.C.L.M.: 39 (1968).—Ross, Fl. Natal: 295 (1972).—Munday & Forbes in Journ. S. Afr. Bot. **45**: 9 (1979).—Compton, Fl. Swaziland: 475 (1976). TAB. **9**. Type from India.
    Ipomoea benguelensis Baker in Kew Bull. **1894**: 69 (1894). Type from Angola.
    Ipomoea phyllosepala Baker in Kew Bull. **1894**: 69 (1894). Type: "Zambesiland", *Kirk* (K, holotype).
    Bonamia volkensii Dammer in Engl., Pflanzenw. Ost-Afr. **C**: 329 (1895). Type from Tanzania.

Stems slender, 1–3 m. long, occasionally rooting at the nodes, more or less pubescent. Leaf lamina very variable, oblong or ovate to broadly ovate, 2·5–14 × 1·2–10 cm., obtuse to acuminate at the apex, cordate, hastate, cuneate or sometimes truncate at the base, dentate or entire, pilose to velvety on both surfaces; auricles entire sometimes angular, occasionally spreading; petiole 16 long, pubescent. Peduncle 0·5–10 cm. long, 1–3 flowered, pubescent; pedicles usually short, up to 3 cm. long; bracts oblong-lanceolate, 0·5–1·7 cm. long. Sepals lanceolate to ovate up to 17 mm. long, more or less hairy and ciliate, the outer 3 much larger than the inner 2, nerved in fruit. Corolla pale yellow or white, usually with a purple or claret centre, 2–3·5 cm. long; limb with 5 very short, rounded, emarginate, mucronulate lobes; midpetaline areas pilose outside. Ovary densely hairy with long white hairs, also with a few long hairs on the basal part of the style. Capsule depressed-globose to more or less quadrangular, crowned by the persistent style, pilose. Seeds 2–4, black, glabrous or nearly so.

Zambia. N: Mulwe Agric. Station, fl. & fr. 12.iii.1969, *Anton-Smith* (SRGH). C: Katondwe, fl. & fr. 4.iv.1966, *Fanshawe* 9663 (SRGH). **Zimbabwe**. W: Hwange, fl. & fr. 11–16.v.1956, *Plowes* (SRGH). E: Chimanimani, 1070 m., fl. & fr. ix.1961, *Goldsmith* 73/61 (LISC; SRGH). **Malawi**. N: Nkhata Bay, 550 m., fl. & fr. 31.iii.1974, *Pawek* 8295 (K; SRGH). C: China area, Nkhota-Nkota, 480 m., fl. & fr. 7.ix.1946, *Brass* 17561 (K). S: Thyolo Mt., 1200 m., fl. & fr. 19.ix.1946, *Brass* 17648 (BM; K; PRE; SRGH). **Mozambique**. N: Moçambique, Angoche, fl. & fr. 22.x.1965, *Mogg* 32352 (LISC; PRE; SRGH). Z: Guruè, 8 km. W. of Gurue, fl. & fr. 7.vii.1942, *Hornby* 4566A (PRE). T: Tete, fl. & fr. vi.1930, *Pomba Guerra* 9 (COI). MS: Chemba, Chiou, fl. & fr. 23.v.1961, *Balsinhas & Marrime* 484 (BM; K; LISC; LMA). GI: Zavala, between Zandamela e Zavala, 3.iv.1959, *Barbosa & Lemos* 8477 (COI; K; LISC; LMA). M: Inhaca Isl., fl. & fr. 27.ix.1958, *Mogg* 28354 (K; LMA; PRE; SRGH).
    Widespread throughout tropical Africa, Asia, Malasia and Polynesia; also in Jamaica as an escape. Rain-forest, mixed open forest, coastal littoral forest and littoral scrub, along dry watercourse, grassland, weed of cultivated land and roadsides; 0–1830 m.

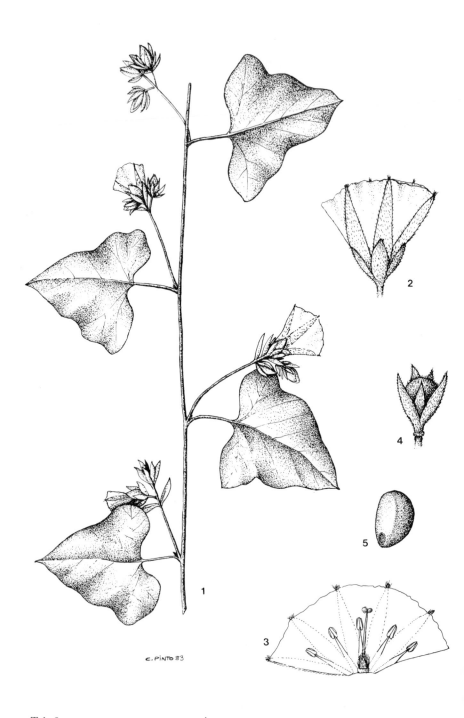

C. PINTO 83

Tab. 9. HEWITTIA SCANDENS. 1, habit ($\times\frac{1}{2}$); 2, flower ($\times$1); 3, corolla opened to show stamens and pistil ($\times$1); 4, fruit ($\times$1); 5, seed ($\times$3). All from *Mogg* 28354.

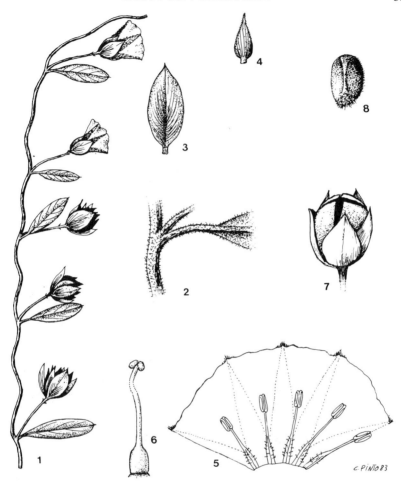

Tab. 10. ANISEIA MARTINICENSIS. 1, habit ($\times\frac{1}{2}$); 2, part of stem, leaf-base and peduncle-base showing the retrorse hairs ($\times 2\frac{1}{2}$); 3, outer sepal ($\times 1$); 4, inner sepal ($\times 1$); 5, corolla opened to show stamens ($\times 1\frac{1}{2}$); 6, pistil ($\times 3$); 7, fruit ($\times 1$); 8, seed ($\times 2\frac{1}{2}$). All from *Balsinhas* 1873.

10. ANISEIA Choisy

Aniseia Choisy in Mem. Soc. Phys. Genève **6**: 481 (1834).

Prostrate or twining herbs. Leaves petiolate, linear, oblong, lanceolate, ovate or elliptic, often mucronulate. Flowers axillary, solitary in the African species, on long peduncles. Sepals 5, herbaceous, acute or acuminate, the 3 outer ones much larger than the inner, ovate and more or less decurrent on the peduncle, enlarged in fruit. Corolla funnel-shaped, usually entire outside, with 5-well-defined hairy midpetaline bands. Stamens 5, included, filiform; pollen smooth. Disk small or absent. Ovary bilocular, 4-ovuled, glabrous; style 1, simple, filiform, included, stigmas, 2, thick, globular or oblong. Capsule ovoid or globose, bilocular, 4-valved and 4-seeded. Seeds trigonous or globose, black.

A small genus of about 5 species occurring throughout the tropics, very close to *Merremia*.

Aniseia martinicensis (Jacq.) Choisy in Mém. Soc. Phys. Genève **8**: 66 (1838).—Hall. f. in Engl., Bot. Jahrb. **18**: 96 (1893).—Dammer in Engl., Pflanzenw. Ost-Afr. **C**: 329

(1895).—Heine in F.W.T.A. ed. 2, **2**: 343 (1963).—Verdc. in F.T.E.A., Convolvulaceae: 48, fig. 13 (1963). TAB. **10**. Types from the West Indies.

Convolvulus martinicensis Jacq., Sel. Stirp. Amer.: 26, t. 17 (1763). Type from Java.

Aniseia uniflora (Burm. f.) Choisy in Mem. Soc. Phys. Genève **6**: 483 (1834); in DC., Prodr. **9**: 431 (1845).—Peter in Engl. & Prantl, Pflanzenfam. **4**, 3a: 25, fig. 12A (1891).—Baker & Rendle in F.T.A. **4**: 88 (1905). Type as above.

Ipomoea lanceolata G. Don., Gen. Syst. **4**: 282 (1837). Type from Sierra Leone.

Aniseia martinicensis (Jacq.) Choisy var. *ambigua* Hall. f. in Engl., Bot. Jahrb. **18**: 96 (1893). Type from Zanzibar.

Twining perennial herb with slender thinly pubescent or glabrescent stems, sometimes rooting below. Leaf lamina narrowly to broadly oblong up to 6 × 2 cm., obtuse at the apex but minutely mucronate, cuneate at base, glabrescent or thinly pubescent, particularly at the margins and on the nerves below; petiole up 10 mm. long. Flowers usually solitary; peduncle up to 3·5 cm. long; pedicels up to 6 mm. long. Sepals ovate up to 15 mm. long, acute, enlarging in fruit up to 23 mm. and becoming scarious and venose. Corolla white, about 2 cm. long. Capsule elongate-ovoid, glabrous. Seeds black, minutely pilose, with a tuft of brown hairs near the hilum.

Mozambique. Z: Maganja da Costa, Bajone, fl. & fr. 18.v.1971, *Balsinhas* 1873 (LMA). Also in Angola, the Congo Republic and West Africa; pantropical. Swampy ground.

11. MERREMIA Dennst.

Merremia Dennst., Schluss. Hort. Malab. 12 et 34 (1818).—Hall. f. in Engl., Bot. Jahrb. **16**: 581 (1893).

Herbs or small shrubs, usually twiners, often prostrate, rarely erect. Stems terete, more rarely winged. Leaves usually petiolate, variable in shape and size, entire, dentate, auricled, lobed or palmately or pedately partite to compound. Flowers axillary, solitary or in few- to many-flowered axillary inflorescences, small to rather large, usually cymose. Bracts usually small. Sepals 5, usually subequal, elliptic to lanceolate, ovate or oblong, acute or obtuse, sometimes accrescent. Corolla funnel-shaped or campanulate, entire or slightly lobed, rarely deeply lobed, mostly with distinct midpetaline areas, white or yellow with various centres, often darker brownish or purplish centre. Stamens often contorted; filaments filiform, often broadened at the base, often unequal; pollen smooth. Ovary 2–4-locular, 4-ovuled; style simple, filifrom, included; stigma biglobose. Disk often annular. Capsule 4-valved or dehiscing irregularly, 1–4-locular. Seeds 4, or less by abortion, glabrous or pubescent.

A genus of about 80 species widely spread in the tropics of both hemispheres and very close to *Ipomoea* differing essentially by having smooth pollen.

1. Leaf lamina entire or 3-lobed; margin crenate or entire; base often sagittate or cordate   2
 — Leaf lamina palmately 3–9-lobed, or palmately compound; lobes entire or pinnatisect; margin entire, crenulate or serrate · · · · · · · · 5
2. Leaf lamina oblong to lanceolate, much longer than broad, with a hastate, sagittate or truncate base · · · · · · · · · 3
 — Leaf lamina ovate with a more or less cordate base · · · · · 4
3. Sepals subequal, up to 10 mm. long, not foliaceous · · · 1. *tridentata*
 — Sepals very unequal, the outer 3 ones up to 20 mm. long, very foliaceous · 2. *medium*
4. Sepals elliptic, acute to obtuse, up to 12 mm. long; capsule with smooth valves · · · · · · · · · · 12. *xanthophylla*
 — Sepals obovate to spathulate, up to 5 mm. long, broadly notched at the mucronate apex; capsule with wrinkled valves · · · · · 13. *hederacea*
5. Leaf lamina pinnately lobed; lobes narrow and linear · · · 3. *pinnata*
 — Leaf lamina palmately lobed or palmately compound · · · 6
6. Stems and peduncles distinctly 4-winged; more or less shrub climber   11. *pterygocaulos*
 — Stems and peduncles not distinctly 4-winged; herb or perennial · · · 7
7. Leaf lamina palmately compound · · · · · · 8
 — Leaf lamina palmately lobed · · · · · · · · 9
8. Leaf lamina 3–5-foliolate; peduncels densley papillate-glandular towards the apex; plant glabrous or with short patent hairs · · · · 5. *quinquefolia*
 — Leaf lamina 5-foliolate; peduncles not glandular; plant covered with long yellow-brown patent hairs up to 6 mm. long · · · · · · 4. *aegyptia*
9. Plant with stellate hairs · · · · · · · · 6. *stellata*
 — Plant without stellate hairs · · · · · · · · 10

10. Corolla 4–5·5 cm. long; sepals 2–3 cm. long - 11
— Corolla 2–3 cm. long; sepals 0·8–1·2 cm. long 12
11. Leaf lamina 5–7 lobed; lobes elliptic-oblong lanceolate, pinnatifid, up to 6·5 × 2 cm.; stems usually muriculate - . . . . 8. *kentrocaulos*
— Leaf lamina 7-lobed; lobes oblong-lanceolate, entire up to 13·5 × 5 cm.; stems smooth . . . . 9. *tuberosa*
12. Corolla up to 2 cm. long; calyx inflated in fruit, enclosing the capsule completely; leaf lamina palmately to pedately 7–11 lobed - . . . . 7. *verecunda*
— Corolla up to 3 cm. long.; calyx not inflated in fruit 13
13. Bracts ovate-lanceolate, subfoliaceous, obscuring the base of the sepals; capsule coriaceous; leaf lamina 3–9 lobed - . . . . 12. *xanthophylla*
— Bracts small, linear, not as above; capsule papyraceous; leaf lamina 5–9 lobed 10. *palmata*

1. **Merremia tridentata** (L.) Hall. f. in Engl., Bot. Jahrb. **16**: 552 (1893); in Engl., Bot. Jahrb. **18**: 116 (1893).—Hiern, Cat. Afr. Pl. Welw. **1**: 729 (1898).—Baker & Rendle in F.T.A. **4**: 111 (1905).—Brenan & al. in Mem. N.Y. Bot. Gard. **9**: 8 (1954).—Macnae & Kalk, Nat. Hist. Inhaca I., Mocambique: 152 (1958).—Heine in F.W.T.A. ed. 2, **2**: 341 (1963). —Verdc. in F.T.E.A., Convolvulaceae: 51, fig. 14 (1963).—Roessler in Merxm., Prodr. Fl. SW. Afr. **116**: 20 (1967).—Binn , H.C.L.M.: 41 (1968).—Jacobsen in Kirkia **9**: 171 (1973).—Compton, Fl. Swaziland: 476 (1976).—Amico in Fitoterapia **48**: 127 (1977); Contr. Con. Fl. Zambezia inf. (Mozamb.) no **49**: 542 (1978).—Munday & Forbes in Journ. S. Afr. Bot. **45**: 9 (1979). Type from India.
    Convolvulus tridentatus L., Sp. Pl.: 157 (1753). Type as above.
    Xenostegia tridentata (L.) Austin & Staples in Brittonia **32**: 533 (1980). Type as above.

A very variable polymorphic perennial. Stems prostrate or twining up to 2 m. long, glabrous or hairy, slender, subterete to angular, striate-ribbed. Leaf lamina linear or lanceolate to oblong, 2–8·5 × 0·1–2·3 cm., more or less obtuse, acuminate to emarginate, mucronulate at the apex, hastate or auriculate at the base, with the lobes often one-several-toothed, glabrous or hairy; petiole 0–3 mm. long. Inflorescence one-few-flowered; peduncle very slender 1–6 cm. long; bracteoles minute, lanceolate or cuspidate, persistent; pedicels thickened upwards, 4–17 mm. long. sometimes winged and crisped. Sepals subequal, oblong or lanceolate, 4–10 mm. long, the outer 2 usually shorter, obtuse or subacute to attenuate-acuminate, cuspidate or mucronate, sometimes winged and crisped. Corolla pale yellow, cream, green or white, often with a darker reddish or brownish centre, funnel-shaped, 10–18 mm. long; limb shallowly 5-lobed with more or less broadly triangular acute lobes and midpetaline areas well defined. Capsule globose to ovoid with papery valves, straw-coloured. Seeds 4 or fewer yellowish brown to dark greyish brown, glabrous.

1. Pedicels and sepals winged - . . . . subsp. *alatipes*
— Pedicels and sepals not winged 2
2. Plants glabrous or glabrescent - . . . subsp. *angustifolia* var. *angustifolia*
— Plants pubescent . . . . subsp. *angustifolia* var. *pubescens*

Subsp. **angustifolia** (Jacq.) van Oststr. in Blumea **3**: 323 (1939).—Meeuse in Bothalia **6**: 706 (1958).—Heine in F.W.T.A. ed. 2, **2**: 341 (1963).—Binns, H.C.L.M.: 41 (1968). —Jacobsen in Kirkia **9**: 171 (1973). Type from Guinea.
    Ipomoea angustifolia Jacq., Ic. Pl. Rar. **2**: 10, t. 317 (1786–93) et Collect. **2**: 367 (1789). Type as above.

Plant glabrous or pubescent. Leaf lamina to lanceolate, usually acute, often thin. Pedicels terete, not winged. Sepals acute, not winged or crisped.

Var. **angustifolia**
    Convolvulus hastatus Desr. in Lam., Encycl. **3**: 542 (1791), non Forsk. (1775). Type from the East Indies.
    Convolvulus filicaulis Vahl, Symb. Bot. **3**: 24 (1794). Type from Guinea.
    Ipomoea denticulata R. Br., Prodr.: 485 (1810). Type from Australia.
    Ipomoea filicaulis (Vahl) Bl., Bijdr.: 721 (1826). Type as above.
    Merremia hastata (Desr.) Hall. f. in Engl., Bot. Jahrb. **16**: 552 (1893).—Dammer in Engl., Pflanzenw. Ost.-Afr. C: 330 (1895). Type as for *Convolvulus hastatus*.
    Merremia angustifolia (Jacq.) Hall. f. in Engl., Bot. Jahrb. **18**: 117 (1893).—Hall. f. in Warb., Kunene-Samb.-Exped. Baum: 345 (1903).—Baker & Rendle in F.T.A. **4**: 111 (1905).—Eyles in Trans. Roy. Soc. S. Afr. **5**: 453 (1916).—Fries, Wiss. Ergebn. Schwed. Rhod.-Kongo-Exped.: 269 (1916).—Hutch. & Dalz., F.W.T.A. **2**: 211 (1931).—Wild, Guide Flora Victoria Falls: 155 (1953). Type from Guinea.
    Merremia angustifolia (Jacq.) Hall. f. var. *ambigua* Hall. f. in Engl., Bot. Jahrb. **18**: 117 (1893).—Hall. f. in Warb., Kunene-Samb.-Exped. Baum: 345 (1903). Type from Ethiopia.

Merremia tridentata (L.) Hall. f. subsp. *hastata* (Desr.) van Ooststr. in Blumea **3**: 317 (1939); Fl. Males., ser. 1, **4**: 445, fig. 27 (1953). Type from the East Indies.

Botswana, N: Maun, fl. & fr. 17.iv.1967, *Lambrecht* 147 (K; LISC; SRGH). SW: Kule, Farm no. 150, 915 m., fl. & fr. 21.ii.1970, *Brown* 8696 (K; PRE; SRGH). SE: 6 km. N. of Morwamosu, fl. & fr.17.ii.1960, *Wild* 5000 (K; SRGH). **Zambia**. B: Kalabo, fl. & fr. 16.xi.1959, *Drummond & Cookson* 6515 (K; LISC; SRGH). N: nr. Nchelenga, L. Mweru, 975 m., fl. & fr. 23.iv.1951. *Bullock* 3823 (K). W: Ndola, Ndola Golf Course, fl. & fr. 22.x.1952, *Angus* 648 (FHO; K). C: Lusaka, nr. Munali Pass, fl. & fr. 19.iv.1957, *Noak* 215 (K; LISC; SRGH). E: Petauke, Kaulu Dam, 900 m., fl. & fr. 3.xii.1958, *Robson* 807 (BM; K; LISC; SRGH). S: Livingstone, 915 m., fl. & fr. vi.1909, *Rogers* 7224 (K; SRGH). **Zimbabwe**. N: Chipuriro, Mwanzamtanda R. area, 550 m., fl. & fr. 30.1.1966, *Müller* 276 (K; SRGH). C: Charter, 1470 m., fl. & fr. xii.1920, *Eyles* 2808 (K; PRE; SRGH). E: Inyanga, 1300 m., *Norlindh & Weimarck* 4349 (COI; PRE; SRGH). S: Chiredzi, Musovi, fl. & fr. 18.v.1971, *Taylor* 182 (K; Lisc; SRGH). **Malawi**. N: Nkhata Bay, Sanga Beach, 490 m., fl. & fr. 22.vi.1969, *Pawek* 2499 (K). C: Nkhota-Kota, 490 m., fl. & fr. 2.ii.1942, *Benson* 665 (K; PRE; SRGH). S: Mangochi town, 370 m., fl. & fr. 24.ii.1979, *Brummitt & Patel* 15452 (K; SRGH). **Mozambique**. N: Mucojo, nr. Macomia, fl. & fr. 30.ix.1948, *Barbosa* 2287 (LISC; LMA). Z: Between Gile and Namiroe R., fl. & fr. 9.x.1949, *Barbosa & Carvalho* 4350 (K; LMA). T: Cahobra Basa, Mucangadzi R., 230 m., fl. & fr. 10.v.1972, *Pereira & Correia* 2496 (LISC; LMA). MS: Manica, Mavita, fl. & fr. 8.xi.1942, *Salbany* 43A (LISC). **GI**: Vilanculos, Bazaruto Is., fl. & fr. 28.x.1958, *Mogg* 28535 (K; LISC; PRE). M: Between Marracuene and Manhica, fl. & fr. 5.vii. 1958, *Barbosa & Lemos* 8289 (COI; K; LMA).

Very common throughout tropical and Southern Africa, tropical Asia to China, Malaysia and Australia. Woodland, savanna, grassland, roadsides, riversides, sandy soils and as weed of cultivation; 0–1550 m.

Var. **pubescens** Rendle in Baker & Rendle, F.T.A. **4**: 112 (1905). Verdc. in F.T.E.A., Convolvulaceae: 52 (1963). Type: Mozambique: Zambesi, *Webb* (BM, holotype).

Caprivi Strip. E of Kwando R., Katima Mulilo, 945 m., fl. & fr. x.1945, *Curson* 1042 (PRE). **Botswana**. N: Moremi Wildlife Reserve, fl. & fr. 22.i.1973, *Smith* 368 (K; SRGH). **Zambia**. B: Sesheke, nr. Sichinga Forest, fl. & fr. 28.xii.1952, *Angus* 1059 (FHO; K). **Zimbabwe**. W: Matobo, Farm Besna Kobila, 1470 m., fl. & fr. xi.1957, *Miller* 4782 (SRGH). C: Makoni Distr., between Rusapi and Nyazura, 1400 m., fl. & fr. 10.xi.1930, *Fries, Norlindh & Weimarck* 2801 (SRGH). **Mozambique**. N: between Corrane and Nampula, fl. & fr. 15.xi.1936, *Torre* 1027 (COI; LISC). Z: Mocuba,Namagoa, 60–120 m., fl. & fr. ix.1943, *Faulkner* 102 (BM; K; PRE). MS: Manica, Revuè Mt., nr. Macequece Rd., fl. & fr. 10.iii.1948, *Garcia* 564 (LISC). M: Maputo, fl. & fr. 2.xii.1894, *Schlechter* 11590 (BM; COI; K; PRE).

Also in Tropical East Africa. Open woodland, grassland, cultivated ground, roadsides, sandy soils; 40–1500 m.

Subsp. **alatipes** (Dammer) Verdc. in Kew Bull. **13**: 186 (1958), F.T.E.A., Convolvulaceae: 52, fig. 14 (1963).—Binns, H.C.L.M.: 41 (1968). Type from Tanzania.
   Merremia alatipes Dammer in Engl., Plflanzenw. Ost.-Afr. **C**: 330 (1895). Type as above.
   Merremia angustifolia (Jacq.) Hall. f. var. *alatipes* (Dammer) Rendle in F.T.A. **4**: 112 (1905). Type as above.

Leaf lamina lanceolate, thin, usually large, glabrous. Pedicels conspicuously winged. Outer sepals with crisped margins. Corolla yellow with a crimson centre, more or less 15 mm. long.

Zambia. E: Msoro, c. 80 km. W. of Chipata in Luangwa Valley, 730 m., fl. & fr. 11.vi.1954, *Robinson* 856 (K). **Zimbabwe**. E: Mutare, Bazeley Bridge Rd., Zimunya Reserve, 915 m., fl. & fr. 16.iii.1958, *Chase* 6852 (K; SRGH). S: S. of Lundi R., Beit Bridge Rd., fl. & fr. 16.ii.1955, *Exell, Mendonça & Wild* 364 (BM; LISC; SRGH). **Malawi**. N: Nkhata Bay Distr., Chinteche, 520 m., fl. & fr. 7.vi.1974, *Pawek* 8681 (K; SRGH). C: Kasungu, 1060 m., fl. & fr. 7.v.1970, *Brummitt* 10440 (K). S: Tangadzi R., c.11 mls. W. of Chiromo, fl. & fr. 16.vii.1958, *Seagrief* 3076 (K; LISC; SRGH). **Mozambique**. N: Nampula, fl. & fr. 2.vi.1935, *Torre* 687 (COI; LISC). Z: Mocuba, Namagoa, fl. & fr. 10.vii.1949, *Faulkner* 448 (K). MS: Manica, 25 km. SW. of Dombe, c.320 m., fl. & fr. 22.iv.1974, *Pope & Müller* 1248 (K; LISC; SRGH). GI: Gaza, between Chiconela and Gumbe, fl. & fr. 26.v.1965, *Pereira, Marques & Balsinhas* 474 (COI; LISC; LMA).

Also in Tropical East Africa. Open forest and evergreen mixed deciduous woodland, grassland, waste places, roadsides and cultivated ground; 300–1100 m.

2. **Merremia medium** (L.) Hall f. in Engl., Bot. Jahrb. **16**: 552 (1893); in Engl., Bot. Jahrb. **18**: 118 (1893).—Verdc. in Kew Bull. **15**: 5 (1961); in F.T.E.A., Convolvulaceae: 52 (1963). Type from "India".

Convolvulus medium L., Sp. Pl.: 156 (1753) excl. syn. Rheede. Type as above.
Aniseia medium (L.) Choisy in Mém. Soc. Phys. Genève 6: 482 (1834). Type as above.
Jacquemontia hastigera Boj., Hort. Maurit.: 229 (1837) *nom. nud.* Type as above.
Xenostegia medium (L.) Austin & Staples in Brittonia **32**: 534 (1980). Type as above.

Twining or prostrate plant very similar in foliage to *M. tridentata*. Stems glabrous, slender, only slightly winged. Leaf lamina lanceolate-elliptic, 2–9·5 × 0·3–2·5 cm., acute or obtuse at the apex, mucronulate, narrowed towards the base and then markedly sagittate-auriculate, glabrous; petiole up to 7 mm. long. Inflorescences 1–2 flowered; peduncle slender, up to 1·4 cm. long, pubescent; bracteoles minute, lanceolate or cuspidate, persistent; pedicels up to 1·4 cm. long, glabrous, slightly winged. Sepals unequal, the outer 3 ovate-oblong, very foliaceous, acute and apiculate at the apex, more or less sagittate with acute lobes at the base, glabrous, venose, in fruit accrescent up to 2 × 0·8 cm.; the inner lanceolate-acuminate, up to 10 × 25 mm. Corolla dark yellow, about 2·3 cm. long. Capsule ovoid, hairy above. Seeds 4 or fewer brown, glabrous.

Mozambique. N: Between Nangade and Mueda, fl. & fr. 18.ix.1948, *Barbosa* 2209 (LMA). Also in Tanzania, Madagascar and the Mascarene Is. Forest and cultivated ground.

3. **Merremia pinnata** (Hochst. ex. Choisy) Hall. f. in Engl., Bot. Jahrb. **16**: 552 (1893); in Engl., Bot. Jahrb. **18**: 116 (1893).—Dammer in Engl., Pflanzenw. Ost.-Afr. **C**: 330 (1895).—Baker & Rendle in F.T.A. **4**: 113 (1905).—Eyles in Trans. Roy. Soc. S. Afr. **1916**: 453 (1916).—Dandy in Andr., Fl. Pl. Anglo-Egypt. Sudan **3**: 126 (1956).—Meeuse in Bothalia **6**: 707 (1958).—Verdc. in F.T.E.A., Convolvulaceae: 55 (1963).—Heine in F.W.T.A. ed. 2, **2**: 341 (1963).—Roessler. in Merxm., Prodr. Fl. SW. Afr. **116**: 20 (1967).—Binns, H.C.L.M.: 41 (1968).—Jacobsen in Kirkia **9**: 171 (1973). Type from Sudan Republic.
   Ipomoea pinnata Hochst. ex Choisy in DC., Prodr. **9**: 353 (1845).—N.E. Br. in Kew Bull. **1909**: 124 (1909). Type as above.

Trailing or twinging annual. Stems many, herbaceous, slender, up to about 70 cm. long, pubescent with soft more or less distinctly bulbous-based spreading hairs (at least when young), as are the leaves, peduncles, bracts, calyces and capsules. Leaves sessile, deeply pinnatifid 1–4·5 × 0·5–2 cm.; lobes in 8–12 pairs, narrowly linear, entire, extending almost to the mid-rib, about 9 × 0·5 mm., the lowest branched and stipule-like. Inflorescences 1–3-(few)-flowered; peduncle up to 4 cm. long; pedicels up to 5 mm. long; bracts small, linear. Sepals unequal, elliptic, 4–8 × 1·5–2 mm., markedly acuminate at the apex, subcoriaceous, pilose with long hairs; inner sepals much smaller than the outer. Corolla white, greenish-white, cream or yellow, up to 7 mm. long, 5-lobed and midpetaline areas not differentiated. Ovary with stiff hairs longer than itself. Capsule globose or ovoid, straw-coloured. Seeds dark brown to black, glabrous, with grey minutely raised markings.

Botswana. N: 77 km. N. of Aha Hills along SW. Africa border trace, fl. & fr. 13.iii.1965, *Wild & Drummond* 6994 (K; SRGH). **Zambia**. B: Masese, fl. & fr. 3.v.1961, *Fanshawe* 6539 (FHO; K). N: Mporokoso, nr. Muzombe, 1040 m., fl. & fr. 15.iv.1961, *Phipps & Vesey-Fitzgerald* 3193 (K; LISC; SRGH). W: Luanshya, fl. & fr. 29.iii.1957, *Fanshawe* 3117 (K; SRGH). C: Kabwe, Chibombo, 1210 m., fl. & fr. 25.ii.1973, *Kornas* 3285 (K) E: Chikwa, c. 112 km. NW. of Lundazi in the Luangwa Valley, 975 m., fl. & fr. 3.vi.1954, *Robinson* 826 (K). S: Namwala, Sibanzi Hill, Lochinvar National Park, fl. & fr. 13.vii.1972, *Lavieren, Sayer & Rees* 825 (K; SRGH). **Zimbabwe**. N: Miami, fl. & fr. 7.iii.1947, *Wild* 1814 (COI; K; SRGH). C: Masvingo, Makaholi Experimental Station, fl. & fr. 10.iii.1978, *Senderayi* 173 (K; SRGH). E: Mutare, Zimunya Reserve, 915 m., fl. & fr. 2.v.1954, *Chase* 5269 (K; LISC; SRGH). S: Buhera, 1150 m., fl. & fr. 20.iv.1969, *Biegel* 2929 (COI; K; SRGH). **Malawi**. N: Rumphi Distr., Livingstonia Escarpment, 564 m., fl. & fr. 20.iv.1972, *Pawek* 5149 (K; SRGH). C: Between L. Malawi and Grand Beach Hotel, nr. Salima, 480 m., fl. & fr. 16.ii.1959, *Robson* 161 (BM; K; LISC; SRGH). S: Mangochi, Monkey Bay, 530 m., fl. & fr. 1.iii.1970, *Brummitt & Eccles* 8815 (K). **Mozambique**. N: Nampula, Mesa Mt., fl. & fr. 13.iv.1961, *Balsinhas & Merrime* 389 (BM; COI; K; LMA; SRGH). Z.: Quelimane, Lugela, Moebede Rd., fl. & fr. 6.iv.1948, *Faulkner* 235 (COI; K; SRGH). T: Tete, Songo, fl. & fr. 8.iii.1972, *Aguiar Macedo* 5024 (COI; K; LISC; LMA). MS: Manica, Revuè Mts. nr. Macequece Rd., fl. & fr. 10.iii.1948, *Garcia* 560 (C; LISC; LMA; MO; WAG).
   Widely spread from the Sudan Republic to West Africa, S. Africa and Namibia. Woodland, mixed savanna woodland, riverine forest, grassland, riverside, sandy dunes, sandy roadside and cultivated ground; 530–1500 m.

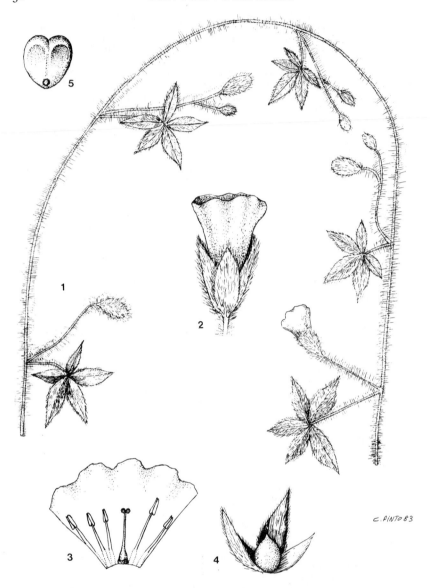

Tab. 11. MERREMIA AEGYPTIA 1. habit ($\times\frac{1}{2}$); 2, flower ($\times$1); 3, corolla opened to show stamens and pistil ($\times$1); 4, fruit after removing two sepals ($\times$1); 5, seed ($\times$3). All from *Pomba Guerra* 8.

4. **Merremia aegyptia** (L.) Urb., Symb. Antill. **4**: 505 (1910).—van Ooststr. in Fl. Males., Ser. 1, **4**: 448 (1953).—Dandy in FW. Andr., Fl. Pl. Anglo-Egypt. Sudan **3**: 125 (1956).—Heine in F.W.T.A. ED. 2, **2**: 341 (1963). TAB. **11**. Type from Tropical Africa.

Ipomoea aegyptia L., Sp. Pl.: 162 (1753). Type as above.
Convolvulus pentaphyllus L., Sp. Pl., ed. 2: 223 (1762). Type from Martinique.
Ipomoea pentaphylla (L.) Jacq., Ic. Pl. Rar. **2**: 10, t. 319 (1786–93); Collect. **2**: 297 (1789). Type as above.
Batatas pentaphylla (L.) Choisy in Mém. Soc. Phys. Genève **6**: 436 (1834); in DC., Prodr. **9**: 339 (1845). Type as above.

Merremia pentaphylla (L.) Hall. f. in Engl., Bot. Jahrb. **16**: 552 (1893); in Engl. Bot. Jahrb. **18**: 115 (1893).—Hiern, Cat. Afr. Pl. Welw. **1**: 728 (1898).—Baker & Rendle in F.T.A. **4**: 108 (1905). Type as above.

Twining sometimes very vigorous herb up to 6 m. high. Stems slender, terete, hirsute with long yellow-brown patent up to 6 mm. long hairs. Leaves palmately compound with 5 leaflets sessile, elliptic or elliptic-oblong, 2·5–9 × 1–4 cm., acute at the base, acute, acuminate at the apex, entire, appressed-pilose on both surfaces; petiole slender, as long as or longer than the lamina, 2·5–11·5 cm. long, patently hirsute. Inflorescences few-several-flowered; peduncle up to 22 cm. long; patently hirsute; pedicels 10–25 mm. long, patently hirsute. Bracts lanceolate, 2–4 mm. long, caducous. Sepals unequal, the 3 outer 14–25 mm. long, ovate-lanceolate, acute or acuminate, densely hirsute; the 2 inner slightly shorter, ovate, acute, glabrous. Corolla funnel-shaped, white, up to 3·5 cm. long, glabrous. Capsule globose, pale brown, glabrous, 4-celled, 4-valved. Seeds 4, glabrous.

Zambia. S: Gwembe, Chief Sikongo's area, 396 m., fl. & fr. 7.vi.1963, *Bainbridge* 802 (FHO; K; SRGH). **Zimbabwe.** N: Kariba, 610 m., fl. & fr. iii.1960, *Goldsmith* 42/60 (K; SRGH). W: Hwange, Deka R., fl. & fr. 21.vi.1934, *Eyles* 7958 (BM; K; SRGH). **Mozambique.** T: Tete, Cahobra Bassa, Zambezi R., c.400 m., fl. & fr. 19.iv.1972, *Pereira & Correia* 2140 (LISC; LMA).
Widespread throughout the tropics. Open forest, open bush-savanna/semi-thicket, grassland, confluence of rivers and riverbanks; 20–610 m.

5. **Merremia quinquefolia** (L.) Hall. f. in Engl., Bot. Jahrb. **16**: 552 (1893).—van Ooststr. in Fl. Males. ser. 1, **4**: 446, fig. 28 (1953).—Verdc. in Kirkia **1**: 29 (1961); in Kew Bull. **33**: 160 (1978). Type from tropical Africa.
  Ipomoea quinquefolia L., Sp. Pl.: 162 (1753). Type as above.
  Convolvulus quinquefolius L., Syst. ed. 10: 923 (1759). Type as above.
  Batatas quinquefolia (L.) Choisy in Mém. Soc. Phys. Genève **8**: 49 (1837); in DC., Prodr. **9**: 339 (1845). Type as above.

Herbaceous or perennial twiner. Stems slender, scandent, glabrous or sparsely to densely hirsute with patent hairs. Leaves palmate, 3—5-foliolate, glabrous; leaflets sessile or shortly petiolulate, lanceolate, narrowly cuneate at the base, acuminate at the apex, margins serrate; central leaflet up to 5·3 × 0·9 cm., lateral leaflets about 2 × 0·4 cm.; petiole 0·5–3 cm. long, glabrous or setose. Inflorescences 1–2(5)-flowered; peduncle up to 5·5 cm. long, densely papillate-glandular towards the apex; pedicels 5–6 mm. long, glabrous or with a few glands near the base; bracts minute, narrow-triangular, caute. Sepals unequal, glabrous, oblong-ovate, subtruncate or emarginate at the apex, mucronulate, the 2 outer smaller, 6–7 mm. long, the inner 3·8–9·5 mm. long. Corolla funnel-shaped, base tubular, yellow, or whitish, 2 cm. long, limb slightly lobed, glabrous. Capsule subglobose, straw-coloured. Seeds black, oblong-ovate in outline, trigonous, covered with a scattered brownish pubescence.

Mozambique. N: Mocambique, Angoche (Antonio Enes), fl. & fr. 22.x.1965, *Mogg* 32347 (K; LISC; SRGH). Z: Mocuba, 75 m., fl. & fr. 5.vi.1949, *Gerstner* 7108 (K; PRE; SRGH). MS: Chimoio Distr., Bandula Mt., 600 m., fl. & fr. 5.iv.1958, *Chase* 6873 (K; SRGH). GI: Caniçado, Guijá, nr. Limpopo R., fl. & fr. 11.vi.1960, *Lemos & Balsinhas* 69 (BM; COI; K; LISC; LMA; SRGH). M: Maputo, Red Point, fl. & fr. 16.vi.1946, *Pedro* 1409 (LMA).
Introduced in Mozambique, also in Tanzania, tropical America and cultivated in Malaysia. Woodland, bush and roadside; 75–600 m.

6. **Merremia stellata** Rendle in Journ. Bot. (London) **46**: 179 (1908).—Verdc. in Kew Bull. **13**: 185 (1958). Type from Angola.
  Astromerremia digitata Pilger in Notizbl. Bot. Gart. Berl. **13**: 107 (1936). Type from Angola.

Herb, trailing up to 1 m. from a woody rootstock. Stem slender, densely pubescent, with stellate hairs. Leaves palmate, mostly 5-foliolate, all parts covered with stellate hairs, more sparsely on the upper surface; leaflets sessile or shortly petiolulate, spreading, oblanceolate, the middle one up to 4·8 × 0·7 cm., the laterals successively smaller, margin slightly undulate; petiole up to 1·2 cm. long. Flowers solitary, axillary; peduncle slender, up to 3 cm. long, densely stellate-pubescent; pedicels slender, up to 2 cm. long, densely stellate-pubescent; bracts minute, lanceolate. Sepals lanceolate, shortly acuminate, about 13 mm. long, the 2 inner slightly shorter, glabrous except for the sparse stellate pubescene on the lower part. Corolla cream-white, up to 2·8 cm. long, with the limb slightly lobed, glabrous. Capsule globose, brown. Seeds blackish, trigonous, pubescent along the angles.

Zambia. W: Kasempa Distr., 48 km. from Chizela on Rd. to Solwezi, fl. & fr. 26.i.1975, *Brummitt, Chisumpa & Polhill* 14143 (K; SRGH).
Also in Angola. Woodland.

7. **Merremia verecunda** Baker & Rendle in F.T.A. **4**: 110 (1905).—N. E. Brown in Kew Bull. **1909**: 123 (1909).—Pilger in Engl., Bot. Jahrb. **48**: 349 (1912).—Meeuse in Fl. Pl. Afr. **30**: pl. 1193 (1955); in Bothalia **6**: 703 (1958). Type: Botswana, Ngamiland, Kgwebe Hills, *Lugard* 134 (BM, lectotype; K, isotype).
    Ipomoea verecunda (Rendle) N. E. Br. in Kew Bull. **1909**: 123 (1909). Type as above..

Trailing herb, about 1·5 m. long. Stems usually several from the base, procumbent, rather slender, glabrous or thinly hairy. Leaves deeply palmately to pedately 7–9 (11)-sect, 2–8 cm. in diam., glabrous above, thinly pubescent beneath; leaflets linear-oblong, narrowly oblong-lanceolate or spathulate-obovate, acute or obtuse, apiculate, tapering at base into the very short portion of the leaf by which they are connected, the lateral ones gradually smaller; petiole up to 5 cm. long. Inflorescence 1-flowered or occasionally cymosely 2–3 flowered; peduncle 5 mm. long; bracteoles linear or linear-lanceolate, acute, thinly pubescent; pedicels varying in length, at first erect or patent, ultimately reflexed. Calyx turbinate, 10–15 mm. long in flower, inflated in fruit; sepals unequal, ovate, subobtuse to acuminate, about 8 mm. long, concave, softly pubescent, with 6–7 purple-brown nerves deeply sulcate-plicate and with small purplish spots, hairy on the nerves, very accrescent in fruit, inner ones shorter and narrower, less concave, without or with only a few purplish stripes, not sulcate or plicate, hardly accrescent in fruit. Corolla widely funnel-shaped, 15–20 mm. long, yellow or whitish with a purplish centre; the limb shallowly 5-lobed to pentagonal, glabrous or thinly pubescent towards the base. Capsule completely enclosed by the inflated calyx, somewhat depressed. Seeds black, shiny, smooth, glabrous except for minute flattened hairs along the ciliate angles.

Botswana. N: Ngamiland 32·5 km. from Matsibe Tsetse Camp, Okavango, 930 m., fl. & fr. 17.iii.1961, *Richards* 14754A (K; SRGH). SW: Ghantzi, 6·4 km. NW. Kang, fl. & fr. 18.ii.1960, *Yalala* 48 (K; SRGH). SE: Kgwena, Gaberones, fl. & fr. 6.v.1955, *Reynecke* 336 (PRE). **Zimbabwe**. W: Plumtree, 1936, *Eyles* 8559 (K).
Also Namibia. On patches of grass in deep dry sand; 930–1070 m.

8. **Merremia kentrocaulos** (C. B. Clarke) Rendle in F.T.A. **4**: 103 (1905).—Hall. f. in Meded. Rijksherb. Leiden **1**: 21 (1910).—Meeuse in Fl. Pl. Afr. **30**: pl. 1194 (1955); in Bothalia **6**: 704 (1958).—Verdc. in F.T.E.A., Convolvulaceae: 59 (1963).—Heine in F.W.T.A. ed. 2, **2**: 342. (1963). Type from Ethiopia.
    Convolvulus kentrocaulos Steud. In Schaed Pl. Schimp. It. Abyss. 2 no. 800 *nom. nud.*
    Ipomoea kentrocaulos in Hook. f., Fl. Brit. Ind. **4**: 213 (1883).—N.E. Br. in Kew Bull. **1909**: 124 incl. var. *pinnatifida* N.E. Br. Type as above.
    Operculina kentrocaulos (C. B. Clarke) Hall. f. in Engl., Bot. Jarhb. **18**: 119 (1893). —Hiern, Cat. Afr. Pl. Welw. **1**: 730 (1898). Type as above.

Large, woody, glabrous, twining perennial. Stems becoming woody, the younger ones slender, herbaceous, terete, usually distinctly muricate with reddish papillae (as are the petioles, the peduncles and the pedicels). Leaf lamina, pentagonal in outline, 4–13 cm., as long and as wide, palmately dissected nearly to the base (base cordate with a narrow-sinus); lobes 5–7, elliptic, lanceolate or oblong-lanceolate, up to 6·5 × 2 cm.; acute, entire or minutely crenulate in var. *kentrocaulos* or deeply pinnatifid in var. *pinnatifida*; petiole 3–7 cm. long, often muriculate. Inflorescence cymose, few-flowered or reduced to a single flower. Peduncle patent to suberect, 3–8 cm. long, often muricate; bracteoles ovate, acute, early deciduous, occasionally larger and dissected like the leaves in var. *pinnatifida*; pedicels up to 3 cm. long at first deflexed, patent to suberect when the flowers open and ultimately cernuous in fruit. Sepals ovate-oblong, up to 3 cm. long, obtuse edges, glabrous, concave, somewhat unequal, enlarging in fruit. Corolla funnel-shaped, yellow or white dark purple centre, glabrous, 3·5–5·5 cm. long; limb faintly petagonal, plicate, lobes bluntly triangular. Capsule narrowly ellipsoid at first enclosed in the accrescent, brown and coriaceous calyx, but ultimately exposed just before dehiscence when sepals spread out, pale brown. Seeds brown to black, minutely hairy.

Var. **pinnatifida** N.E. Br. in Kew Bull. **1909**: 124 (1909). Type: Botswana, Kgwebe Hills, *Lugard* 82 (K, holotype).

Botswana. N: Ngamiland, c.5 km. NW. of Kgwebe Hills, fl. & fr. 15.ii.1966, *Drummond* 8730 (SRGH). SW: near Farm 115, fl. 21.ii.1970, *Brown* 8743 (K). SE: 3 km. N. of Shashi, fl. 21.i.1960, *Leach & Noel* 286 (K). **Zimbabwe.** N: Darwin, W. of Sanga R., 608 m., fl. 28.i.1960, *Phipps* 2464 (K). W: Bulalima-Mangwe, fl. & fr. 10.v.1942, *Feiertag* 45476 (SRGH). C: Marondera, Odzi R., fl. & fr. xi.1931, *Myres* 615 (K). E: Chimanimani, 488 m., fl. & fr. 17.xii.1952, *Chase* 4743 (BM; K; SRGH). S: Gwanda, 608 m., fr. 10.v.1958, *Drummond* 5738 (K; LISC; SRGH). **Mozambique.** T: Tete, nr. Changara, Tete-Harare Rd., 450 m. fl. & fr. 1.iii.1961, *Richards* 14507 (K; LISC). MS: Ancueza Distr., Chiou, fl. & fr. 12.iv.1960, *Lemos & Macuacua* 85 (COI; K; LISC; LMA). GI: Canicado, nr. Mabalane, fl. & fr. 4.vi.1959, *Barbosa & Lemos* 8608 (COI; K; LISC; LMA; SRGH).

Also in S.Africa and India. Woodland, open forest scrubland, grassland, sandy and rocky soils; 200–1300 m.

The var. *kentrocaulos* has leaf lobes entire or crenate, not pinnatifid. It occurs in Nigeria, Sudan, Ethiopia, Uganda and Angola.

9. **Merremia tuberosa** (L.) Rendle F.T.A. **4**: 104 (1905).—van Udststr. in Fl. Males., Ser. 1, **4**: 447, fig. 29 (1953).—Verdc. in F.T.E.A., Convolvulaceae: 60 (1963). Types from Jamaica?
    Ipomoea tuberosa L., Sp. Pl.: 160 (1753). Type as above.
    Ipomoea mendesii Welw., Apont.: 584 (1859). Type from Angola.
    Operculina tuberosa (L.) Meisn. in Mart., Fl. Brasil. **7**: 212 (1869).—Hall. f. in Engl., Bot. Jahrb. **18**: 119 (1893).—Hiern, Cat. Afr. Pl. Welw. 1; 730 (1898). Type from Angola.
    Merremia kentrocaulos sensu Brenan, T.T.C.L.: 172 (1949), non (C. B. Clarke) Rendle 1905.

A glabrous perennial twiner or liane. Stems smooth, from a large subterranean tuber, robust, terete and finely striate. Leaf lamina circular 6–16 × 6–16 cm., palmately 7-lobed to near the base; segments oblong-lanceolate or oblanceolate, up to 13·5 × 5 cm., acuminate at the apex, narrowed below, entire, the middle segment larger than the lateral ones; petiole 6–15 cm. long, slender. Inflorescence few- to several-flowered. Peduncle terete or more or less applanate towards the apex, 4–13 cm. long; pedicels 1·5–1·8 cm. long, enlarging in fruit to 5 cm.; bracts triangular, more or less 2 mm. long, two outer sepals ovate to broadly ovate, 21–25 mm. long, obtuse, indistinctly minutely mucronate; the three inner ones narrower, oblong, 20–23 mm.; all sepals enlarging in fruit to 5–6 mm. long, enclosing the capsule. Corolla funnel-shaped with a more or less cylindrical tube, yellow, up to 5·5 cm. long, glabrous. Capsule ellipsoid to globose, enclosed by the sepals, splitting irregularly and also separating about the base. Seeds black, pubescent, with somewhat longer black hairs along the margins.

Mozambique. MS: Dondo to Vila Machado, 912 m., fl. & fr. 24.iii.1955, *Chase* 5536 (BM; K; SRGH). M: Maputo, Red Point, fr. 16.vi.1946, *Pedro* 1413 (LMA).

Native of tropical America, often cultivated and sometimes naturalised in tropical East Africa, Angola, the Mascarene Is., India, Sri Lanka and Malaysia. Sandy soils; 915 m.

10. **Merremia palmata** Hall. f. in Engl., Bot. Jarhb. **18**: 112 (1893).—Baker & Rendle in F.T.A. **4**: 108 (1905).—Eyles in Trans. Roy. Soc. S. Afr. **5**: 453 (1916).—Meeuse in Fl. Pl. Afr.: **32**, t. 1245 (1957); in Bothalia **6**: 704 (1958).—Verdc. in F.T.E.A., Convolvulaceae: 58 (1963).—Roessler in Merxm. in Prodr. Fl. SW. Afr. **116**: 20 (1967).—Compton, Fl. Swaziland: 475 (1976). Type from Namibia.

Glabrous, prostrate or occasionally twining perennial herb. Stems herbaceous up to 2·5 m. long, sulcate and/or ribbed to almost winged. Leaf lamina deeply palmately 5–7 (9)-lobed (lower most lobes sometimes unequally forked); lobes narrowly linear to lanceolate, acute or obtuse, mucronulate, 0·8–7 × 0·3–1·4 cm.; petiole 1·2–2·5 cm. long, slender. Inflorescence solitary or occasionally cymosely 2–3 flowered; peduncle 1–6·5 cm. long, slender; bracts small, linear; pedicels 0·5–2·5 cm. long, somewhat thicker upwards. Sepals subequal, pale yellowish green, elliptic with rounded apex, glabrous, coriaceous with a membranous edge up to 10 mm. long. Corolla pale yellow or sulphur-yellow with a deep red, maroon or deep magenta centre, up to 3 cm. long, broadly funnel-shaped; limb faintly lobed, spreading; midpetaline areas sparsely pubescent towards the apex. Capsule usually distinctly exserted from the calyx, globose or ovoid-conical, glabrous, pale yellowish green turning straw-coloured, valves papyraceous. Seeds dark greyish brown to black, rather dull, glabrous, nearly smooth to distinctly rugose.

Botswana. N: Xaudum Valley, 16 km. W. of Nxau Nxau, fl. & fr. 14.iii.1965, *Wild & Drummond* 7008 (K; LISC; SRGH). SE: 1·6 km. N. of Maope, fl. & fr. 21.i.1960, *Leach* & Noel 282 (K; SRGH). **Zimbabwe.** W: Bulalima-Mangwe, 1100 m., fl. & fr. 22.xii.1971, Norgrann 78 (K; LISC; SRGH). C: Gweru, Mlezu School Farm, c. 29 km. S.SE. of Kwekwe 1280 m., fl. & fr. 9.i.1966, *Biegel* 778 (K; LISC; SRGH). E: Inyanga, Cheshire, 1300 m., fl. & fr. 15.i.1931, **Norlindh & Weimarck** 4328 (K; PRE; SRGH). S: c. 48–96 km. S. of Masvingo, fl. & fr. 16.iv.1948, *Rodin* 4239 (K; PRE). **Mozambique** N: Cabo Delgado, between Montepuez and Nantulo, 500 m., fl. & fr. 27.xii.1963, *Torre & Paiva* 9743 (LISC; SRGH; UPS). GI: Gaza, Massingir, fl. & fr. 12.ii.1973, *Lousã & Rosa* 304 (LMA). **M**: Maputo, between Umbeluzi and Namaacha, fl. & fr. 16.x.1940, *Torre* 1776 (LISC).

Tropical East Africa, Angola, Namibia, S. Africa (Transvaal) and also India. Open dry deciduous bushland or woodland, roadsides, in sandy soils; 500–1370 m.

11. **Merremia pterygocaulos** (Steud. ex Choisy) Hall. f. in Engl., Bot. Jahrb. **16**: 552 (1893); in Engl., Bot. Jahrb. **18**: 113 (1893).—Dammer in Engl., Pflanzenw. Ost.-Afr. **C**: 330 (1895).— Hiern. Cat. Afr. Pl. Welw. **1**: 727 (1898).—Baker & Rendle in F.T.A. **4**: 105 (1905).—Engl. in Sitz.—Ber. Königl. Preuss. Akad. Wiss. Berl. **52**: 14 (1907).—Eyles in Trans. Roy. Soc. S. Afr. **5**: 453 (1916).—Fries, Wiss. Ergebn. Schwed. Rhod.-Kongo-Exped. **1**: 268 (1916).—Wild, Guide Fl. Victoria Falls: 155 (1953).—Dandy in F. W. Andr., Fl. Pl. Anglo Egypt. Sudan **3**: 124 (1956).—Meeuse in Bothalia **6**: 702 (1958). —White, F.F.N.R.: 363 (1962).—Verdc. in F.T.E.A., Convolvulaceae: 57 (1963). —Heine in F.W.T.A. ed. 2, **2**: 342 (1963).—Jacobsen in Kirkia **9**:171 (1973).—Compton, Fl. Swaziland: 476 (1976). Type from Ethiopia.

Convolvulus pterygocaulos Steud., in Schaed. Schimper, Pl. Abyss. 2, no. 630.

Ipomoea pterygocaulos Choisy in DC., Prodr. **9**: 381 (1845). Type as above.

Ipomoea quadrangularis Steud. ex Choisy in DC., Prodr. **9**: 387 (1845). Type from Madagascar.

Ipomoea petersiana Klotzsch in Peters, Reise Mossamb., Bot.: 239, t. 38 (1861). Type: Mozambique, Sena, *Peters* (B, holotype†).

Perennial, more or less shrubby climber. Stems twining, almost glabrous, 4-winged with wings up to 2 mm. wide, ultimate branches slender, 4-angled. Leaf lamina ovate-circular, cordate at the base, up to 8 × 10 cm., palmately 3–7 lobed to the middle or beyond; lobes elliptic, ovate or oblong, acute or cuspidate, more or less mucronate, entire or subrepand; petiole narrowly or widely winged up to 8·5 cm. long. Inflorescence cymosely few-flowered, rarely solitary; peduncle winged, up to 12 cm. long; bracteoles minute linear; pedicels up to about 3 cm. long, often conspicuously scarred, thickened and remaining erect in fruit. Sepals ovate-oblong or oblong, usually obtuse, up to 10 mm. long, chartaceous, accrescent, becoming broadly ovate to circular and ultimately spreading in fruit. Corolla broadly funnel-shaped, pale yellow, cream or white with red or purple throat, 3·5 cm. long; midpetaline areas strugosely pilose above. Capsule ovoidconical more or less truncate or flattened-depressed at the apex and crowned with the persistent style-base, brown, glabrous. Seeds subglobose, black, smooth and glabrous.

Botswana. N: Kwando R., fl. & fr. 13.xi.1980, *Smith* 3569A (K). **Zambia**. N: Mbala Distr., Mbala-Kambole Rd., 1500 m., fl. 10.ix.1960, *Richards* 13207 (K). W: Mufulira, fl. & fr. 28.v.1934, *Eyles* 8194 (BM; K). C: Chilanga, 1100 m., fl. & fr. x.1929, *Sandwith* 140 (K). E: c. 51 km. from Chipata to Lundazi, fl. & fr. 24.ii.1971, *Anton Smith* 213 (SRGH). S: Gwembe Valley, Nangombe R., nr. Sinazese, fl. & fr. 17.vi.1961, *Angus* 2909 (FHO; K; SRGH). **Zimbabwe.** N: Lomagundi, Rukuti Farm, Doma area, 1160 m., fl. & fr. 1.ix.1963, *Jacobsen* 2197 (PRE). W: Hwange, 920 m., fl. & fr. 26.iii.1974, *Gonde* 85/74 (K; SRGH). E: Inyanga, Pungwe Valley, 1830 m., fl. & fr. 17.vii.1948, *Chase* 900 (BM; K; SRGH). **Malawi**. N: Nkahata Bay, Chombe Estate, fl. & fr. 10.ix.1955, *Jackson* 755 (K; SRGH). S: Mangochi Distr., Namwera, 855 m., fl. & fr. 22.viii.1976, *Pawek* 11646 (SRGH). **Mozambique.** N: Maniamba, Lunha R., fl. & fr. x. 1964, *Magalhães* 19 (COI). Z: c. 12·8 km. W. of Gúruè, *Hornby* 4566B (PRE). T: Tete, Zóbuè, fl. & fr. 27.viii.1943, *Torre* 5797 (C; LISC; WAG). MS: Búzi, Búzi R. Valley, 608 m., fl. & fr. vi.1964, *Goldsmith* 34/64 (K; LISC; SRGH). M: between Vila Luisa and Mahiça, fl. & fr. 1.vii.1949, *Myre* 734 (LMA; SRGH).

Widely spread practically throughout the whole of Africa south of the Sahara, from Ethiopia and the Sudan to W. Africa and S. Africa, also in Madagascar. Forest, mixed woodland, grassland, sandy alluvial soil and roadside; 430–1830 m.

12. **Merremia xanthophylla** Hall. f. in Engl., Bot. Jahrb. **16**: 552 (1893); in Engl., Bot. Jahrb. **18**: 113 (1893).—Baker & Rendle in F.T.A. **4**: 111 (1905).—Dandy in F. W. Andr., Fl. Pl. Anglo-Egypt. Sudan **3**: 125 (1956).—Verdc. in Kew Bull. **33**: 161 (1978). Types from Ethiopia.

    Merremia pes-draconis Hall. f. in Bull. Herb. Boiss. **6**: 537 (1898).—Baker & Rendle in F.T.A. **4**: 107 (1905). Type from Zaire.

A polymorphic herb. Stems prostrate or trailing, subcompressed and angular, clothed, as is the whole plant, with short yellow-brown pubescence. Leaf lamina very variable, from lanceolate-ovate, acuminate, entire to deeply 3–9 lobed, up to 9 × 8 cm., with a shallowly cordate base; lobes linear to lanceolate, more or less entire, acute or obtuse, up tro 6 × 1·5 cm., the terminal one generally much longer than the lowest ones, minutely hirsute on both surfaces with prominent veins on the inferior; petiole 0·3–0·6 cm. long. Inflorescence cymosely few-flowered, more or less dense; peduncle up to 11 cm.; bracts ovate-lanceolate to lanceolate, acute, up to 12 mm. long, subfoliaceous, obscure at the base of the sepals; pedicels 2–6 mm. long. Sepals elliptic, up to 12 mm. long, acute to obtuse, pubescent, coriaceous. Corolla funnel-shaped pale yellow or whitish up to 3 cm. long; midpetaline areas very strigosely pilose, above all at the apices. Capsule yellow-brown, coriaceous laciniate when dehiscent. Seeds not seen.

    Zambia. E: Chipata, nr. Lupande R., 610 m., fl. 1930, *Bush* 17 (K). **Zimbabwe.** N: Darwin, c.18 km. E-NE. of Chipuriro, fl. & fr. 16.iii.1960, *Drummond* 6842 (K; SRGH). Chater, Mhlaba Hills, nr. Windsor, fl. & fr. 16.i.1962, *Wild* 5605 (K; SRGH). E: Mutare, SE. Commonage, 1100 m., fl. & fr. 16.xii.1954, *Chase* 5362 (BM; K; LISC; SRGH).

    Also in Ethiopia, the Sudan Republic and Tropical Africa. Grassland and stony ground; 980–1660 m.

13. **Merremia hederacea** (Burm. f.) Hall. f. in Engl., Bot. Jahrb. **18**: 118 (1893).—van Ooststr. in Fl. Males., ser. 1, **4**: 441, fig. 23 a–b (1953).—Dandy in F. W. Andr., Fl. Pl. Anglo-Egypt. Sudan **3**: 126 (1956).—Meeuse in Bothalia **6**: 700 (1958).—Verdc. in F.T.E.A., Convolvulaceae: 54 (1963).—Heine in F.W.T.A. ed. 2, **2**: 341 (1963). Type from the E. Indies.

    Evolvulus hederaceus Burm. f., Fl. Ind.: 77, t. 30, fig. 2 (1768). Type as above.

    Merremia convolvulacea Hall. f. in Engl., Bot. Jahrb. **16**: 552 (1893). Hiern, Cat. Afr. Pl. Welw. **1**: 729 (1898).—Baker & Rendle in F.T.A. **4**: 114 (1905). Type from India.

A twining or prostrate herb. Stems slender, glabrous or pubescent, smooth or minutely tuberculate, sometimes rooting at the nodes. Leaf lamina ovate, 1·5–5 × 1·2–3 cm., broadly crenate, or shallowly to deeply 3-lobed, glabrous or sparsely hairy; petiole 0·5–3 cm. long, slender, sparsely tuberculate, especially in the basal half. Flowers solitary or several in lax, branched inflorescences; peduncle 1–7 cm. long, thicker than the petioles, usually glabrous or minutely tuberculate; pedicels 1·2–8 mm., similar; bracts narrow-obovate, mucronulate, 3 mm. long, caducous. Sepals obovate to spathulate, glabrous or occasionally slightly pilose on the reverse along the margins, concave, broadly notched at the apex, distinctly mucronulate, outer ones 3.5–4 mm. long, inner ones up to 5 mm. long. Corolla yellow or white, campanulate, 5–12 mm. long, glabrous outside, pilose inside with long hairs near the hairy base of the filaments. Capsule globose or broadly conical, obscurely 4-angled; valves reticulately wrinkled. Seeds more or less pubescent.

    Zambia. S: Kazungula, fl. & fr. ii. 1911, *Gairdner* 542 (K). **Zimbabwe.** N: Urungwe, Mana Pools Game Reserve, fl. & fr. viii.1970, *Guy* 1038 (SRGH). **Mozambique.** T: Tete, Boroma, fl. & fr. 26.vii.1950, *Chase* 2844 (BM; K; SRGH). MS: Gorongosa National Park, 40 m., fl. & fr. 5.v.1964, *Torre & Paiva* 12241 (C; LISC; LMA; WAG).

    Tropical Africa, Madagascar, tropical Asia to China, Malaysia, Queensland and some Pacific islands. Riverine forest, grassland, sandy soils; 40–275 m.

12. OPERCULINA S. Manso

Operculina S. Manso, Enum. Subst. Bras.: 16 (1836).

Large herbaceous twiners. Stems, peduncles and petioles often winged. Leaves petioled; lamina entire, angular often cordate at the base. Flowers usually large, in one to few flowered axillary cymes. Bracts often large, caducous. Sepals 5, large, pergameneous to coriaceous, mostly glabrous, often ventricose, often enlarged in fruit and ultimately with an irregularly lacerate margin. Corolla broadly funnel-shaped or campanulate, white or yellow, glabrous or with the midpetaline bands hairy outside. Stamens 5, included, anthers large, often longitudinally twisted; pollen ellipsoid, smooth. Ovary bilocular, each loculus 2-ovuled; style simple; stigma biglobular. Capsule large, the outer layer splitting around the middle, the upper or lid separating from the lower part, and the inner layer bursting irregularly. Seeds up to 4, pilose or glabrous, blackish.

A genus of about 15 species occurring throughout the tropics. Very close to *Merremia* but well characterised by the differently dehiscent fruit.

Operculina turpethum (L.) S. Manso, Enum. Subst. Bras.: 16 (1836).—van Ooststr. in Fl. Males., ser. 1, 4, **4**: 456, fig. 32 a–b (1953).—Meeuse in Bothalia **6**: 708 (1958).—Verdc. in F.T.E.A., Convolvulaceae: 61 fig. 15 (1963). TAB. **12**. Type from Sri Lanka.
  Convolvulus turpethum L., Sp. Pl.: 155 (1753). Type as above.
  Ipomoea turpethum (L.) R. Br., Prodr.: 485 (1810). Type as above.
  Operculina turpethum Peter in Engl. & Prantl, Nat. Pflanzenfam. ed. 1, **4**, 3a: 32 (1891). Type as above.
  Ipomoea diplocalyx Baker in Kew Bull. **1894**: 71 (1894). Type: Mozambique, Zambezi Delta, Vicenti, *W. L. Scott* (K, holotype).
  Merremia turpethum (L.) Rendle in F.T.A. **4**, 2: 102 (1905). Type as for *Operculina turpethum*.

Perennial twiner with long much branched roots. Stems slender, narrowly 3–5 winged, angular or grooved, glabrous or shortly pilose, often rooting in water; young parts sometimes more or less tomentose. Leaf lamina very variable in shape, circular to lanceolate, 4–10 × 1–9 cm., acuminate to obtuse at the apex, mucronulate, cordate, glabrous to velvety; petiole up to 5 cm. long, terete or sometimes winged. Inflorescences cymosely 1-few flowered; peducle 2–10 cm. long terete or sometimes winged like the stems, glabrous or pubescent, pedicels up to 3 cm. long, enlarging in fruit and becoming clavate, bracts large, oblong or elliptic 1·5–2·5 × 0·5–1 cm., caducous. Sepals ovate or broadly ovate, acute or acuminate, outer ones up to 2·5 × 1 cm., pubescent, inner ones about 2 cm. long, glabrescent, all accrescent in fruit, the calyx becoming broadly cupshaped, up to 6 cm. in diam. Corolla white, pinkish or yellowish, broadly funnel-shaped, 3–4·5 cm. long. Capsule depressed-globose about 1·5 cm. in diam., with circumscissile epicarp, the upper part of which comes off as an operculum or lid. Seeds 4 or fewer, black, glabrous.

Mozambique. Z: Chinde, delta of the Zambezi R., nr. Vicenti, fl. 29.ix.1887, *W. L. Scott* (K).
  Tropical E. Africa, tropical Asia, Australia, Polynesia and the Mascarene Islands, introduced into the W. Indies. Waste places, swamps.

13. LEPISTEMON Blume

Lepistemon Blume, Bijdr. Fl. Neerland Ind.: 722 (1826).

Herbaceous or woody twiners, usually hairy. Leaves petioled; lamina broad, ovate to circular, entire or lobed, often cordate at the base. Flowers in dense, axillary, sessile or peduncled cymes. Sepals 5, subequal, herbaceous or subcoriaceous, acute or obtuse, hairy or glabrous. Corolla urceolate, the tube narrowing upwards; limb shortly 5-lobed. Stamens inserted low down in the corolla tube, dilated in their basal portions into large concave scales which arch over the ovary; pollen spinulose. Ovary glabrous or hairy, bilocular, each loculus with 2 ovules; style 1, very short; stigmas 2, capitate, or one, bilobed. Capsule 4-valved or almost indehiscent. Seeds 4.

A genus of about 10 species in the tropics of the Old World.

Lepistemon owariense (Beauv.) Hall. f. in De Wild., Ann. Mus. Congo, Ser. Bot. **4**: 112

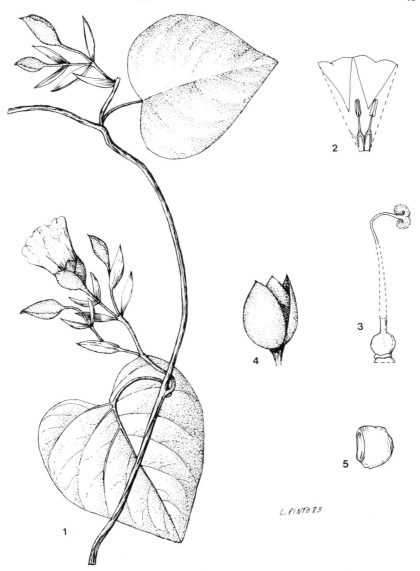

Tab. 12. OPERCULINA TURPETHUM. 1, habit ($\times\frac{1}{2}$); 2, part of corolla, opened to show two stamens ($\times$1); 3, pistil ($\times$4); 4, young fruit, enclosed in calyx ($\times$1); 5, seed ($\times$2). All from *Carvalho* s.n.

(1903).—Dandy in F. W. Andr., Fl. Pl. Anglo-Egypt. Sudan **3**: 123 (1956).—Verdc. in F.T.E.A., Convolvulaceae: 63, fig. 16 (1963).—Heine in F.W.T.A. ed. 2, **2**: 343 (1963). —Binns, H.C.L.M.: 40 (1968). TAB. **13**. Type from Nigeria.

Ipomoea owariensis Beauv., Fl. Owar. **2**: 41, t. 82 (1816). Type as above.

Lepistemon africanum Oliv. in Hook., Ic. Pl. **13**: t. 1270 (1878).—Hall. f. in Engl., Bot. Jahrb. **18**: 123 (1893).—Dammer in Engl. Pflanzenw. Ost.-Afr. **C**: 331 (1895).—Hiern, Cat. Afr. Pl. Welw. **1**: 731 (1898).—Hall. f. in Engl., Bot. Jahrb. **28**: 30 (1899).—Baker & Rendle in F.T.A. **4**: 115 (1905). Type: Mozambique, Zambezi, *Kirk* (K, syntype) and the Sudan, Jur, *Schweinfurth* 1430 (K, syntype).

Ipomoea repandula Baker in Kew Bull. **1895**: 113 (1895). Type from Nigeria.

Lepistemon lingnosum Dammer in Engl., Pflanzenw. Ost.-Afr. **C**: 331 (1895). Type from Tanzania.

Climbing perennial, up to 3 m. Stems covered with long appressed yellow-brown bristly hairs. Leaf lamina cordate-ovate, up to 15 cm. long and wide, acute to emarginate at the apex, entire or shallowly lobed or very coarsely dentate, with deltoid lobes about 1 × 2 cm., pilose on both surfaces; petiole up to 11 cm. long, hairy like the stem. Inflorescences cymosely many-flowered, sessile or shortly peduncled; pedicels up to 2 cm. long, usually hairy; bracts minute, calyx open; sepals ovate to elliptic 6 ×3·5 mm., obtuse, more or less cuspidate, pilose with long yellow hairs to glabrous. Corolla white, up to 1·8 cm. long; tube at first subcylindric, subcampanulate, becoming avoid as the ovary expands; limb up to 3 cm. in diam. Capsule ovoid-globose, setose except in the upper part, indehiscent or bursting irregularly, coriaceous. Seeds grey-black, glabrous, minutely shallowly pitted, globose or nearly so.

Zambia. C: Mpongwe, fl. & fr. 30.iii.1964, *Fanshawe* 8399 (K; SRGH). E: Lunkwakwa, Chipata, fl. & fr. 20.ix.1966, *Mutimushi* 1442 (K; SRGH). **Zimbabwe**. E: Mutare, Victory Ave., 110 m., fl. & fr. 4.iv.1961, *Chase* 7453 (K; LISC; SRGH). S: Chibi, Lundi R., fl. & fr. 2.v.1962, *Drummond* 7848 (K; LISC; SRGH). **Malawi. N**: Chitipa Distr., Kaseye Mission, c. 16 km. E. of Chitipa, 1250 m., fl. & fr. 25.iv.1977, *Pawek* 12643 (SRGH). C: Dedza Mt., fl. & fr. 25.iii.1969, *Salubeni* 1286 (K; SRGH). S: Zomba, fl. & fr. 28.v.1950, Wiehe 557 (K; SRGH). **Mozambique.** Z: Morrumbala, Shire R., 1863, *Kirk* (K). T: Angónia, Dómue Mt., 1400 m., fl. & fr. 9.iii.1964 *Torre & Paiva* 11070 (C; LISC; MA; WAG). MS: Chimoio, Bandula, 700 m., fl. & fr. 30.iii.1952, *Chase* 4447 (BM; K; SRGH).

From the Sudan to Angola. Woodland, thicket, savanna woodland, riverine forest, edge of cultivation and waste ground; 1100–1370 m.

14. MINA Cerv.

Mina Cerv. in La Llave & Lex., Nov. Veg. Descr. **1**: 3 (1824).

Herbaceous twiner. Leaves petioled, entire or mostly palmately lobed. Flowers in axillary, peduncled few to several-flowered monochasial cymes, often secund, pedicels short; bracts minute. Sepals 5, subequal, herbaceous, distinctly aristate, not enlarged in fruit. Corolla narrowly tubular at the base, tubular or urceolate above, slightly curved, with a slightly constricted 5-toothed mouth. Stamens long-exserted; filaments inserted at the apex of the narrow basal tubular part of the corolla; pollen globular, spinulose. Disk annular. Ovary glabrous, 4-locular, each loculus with one ovule; style 1, long-exserted, filiform; stigma bicapitate. Capsule 4-locular, 4-valved. Seeds 4 or less, glabrous.

A monotypic genus which has usually been considered a part of *Ipomoea* sect. *Quamoclit*, but distinct by the curved corolla and long-exserted sexual organs.

Mina lobata Cerv. in La Llave & Lex., Nov. Veg. Descr. **1**: 3 (1824).—Peter in Engl. & Prantl, Pflanzenfam. ed. 1, **4**, 3a: 26 (1891).—van Ooststr. in Blumea **5**: 339 (1943); Fl. Males., ser. 1, **4**: 408 (1953).—Verdc. in F.T.E.A., Convolvulaceae: 68, fig. 18 (1903). TAB. **14**. Type from Mexico.
  Quamoclit lobata (Cerv.) House in Bull. Torr. Bot. Club. **36**: 602 (1909). Type as above.
  Ipomoea lobata (Cerv.) Thell. in Viertel. Nat. Ges. Zürich **64**: 775 (1919).—O'Donell in Bol. Soc. Arg. Bot. **6**: 182 (1957). Type as above.

Perennial glabrous twiner. Stems slender, terete up to 5 m. long. Leaf lamina broadly ovate to reniform, 6–15 cm. long and wide, with a deep sinus at the base, entire or 3-lobed; central lobe elliptic, acuminate, entire or remotely toothed; lateral lobes subovate, often lobed or with coarse teeth; petiole 3–10 cm. long. Inflorescences up to 15 cm. long; peduncles up to 20 cm. long. Sepals 2–4 mm. long, with an awn 2–3 mm. long. Corolla red, later becoming whitish or pale yellow; narrow tubular part about 5 mm. long; upper part 17–20 mm. long; teeth mucronulate. Stamens and style finally twice as long as the corolla; filaments pubescent towards the base. Disk shallowly 5-lobed. Capsule ovoid, glabrous.

Zimbabwe. E: Chimanimani commonage, fl. 1.iv.1980, *Bullock* in GHS 265912 (SRGH).
Also in Mexico to S. America; widely cultivated in the tropics and temperate regions and occasionally seen as an escape.

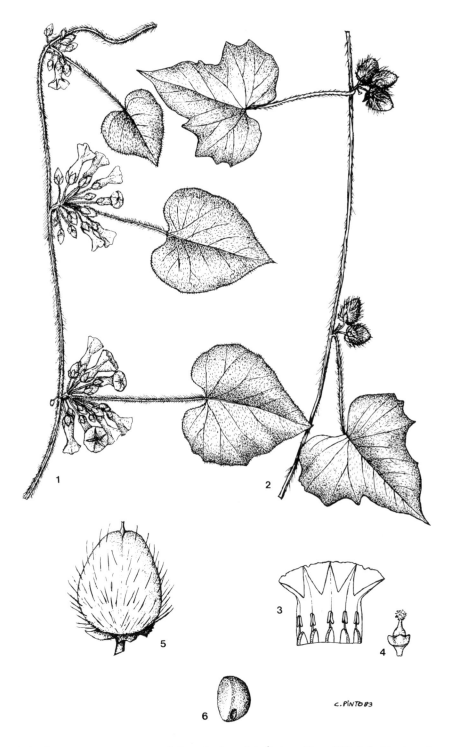

Tab. 13. LEPISTEMON OWARIENSIS. 1, flowering branch ($\times\frac{1}{2}$), from Chase 5215; 2, fruiting branch ($\times\frac{1}{2}$), *Brummitt* 11675; 3, corolla opened to show stamens ($\times 1$); 4, pistil with nectary-disk ($\times 1$), 3–4 from *Chase* 5215; 5, fruit ($\times 2$); 6, seed ($\times 3$). 5–6 from *Brummitt* 11675.

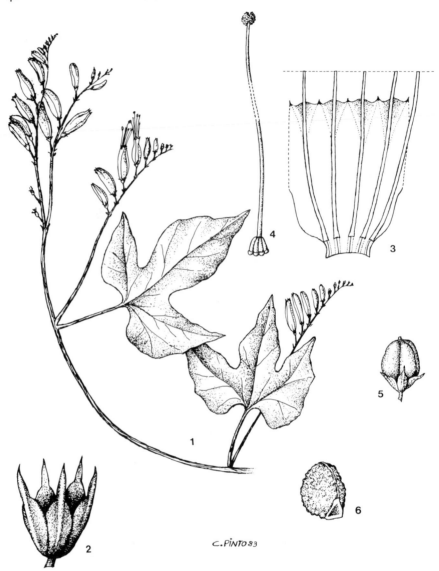

Tab. 14. MINA LOBATA. 1, habit ($\times\frac{1}{2}$); 2, calyx ($\times$4); 3, corolla opened to show part of filaments ($\times$4); 4, pistil ($\times$4); 5, fruit ($\times$2); 6, seed ($\times$4). All from *Bullock* in GHS 265912.

15. STICTOCARDIA Hall. f.

Stictocardia Hall. f. in Engl., Bot. Jahrb. **18**: 159 (1893).—Verdc. in Kew Bull. **15**: (1961).

Herbaceous or woody twiners. Leaves entire, ovate to circular, cordate at the base, densely covered with minute glands beneath (appearing as black dots in dried specimens). Flowers in axillary, peduncled, 1-many-flowered cymes; bracts small, deciduous. Sepals 5, subequal, ovate to circular, subacute or more usually blunt, truncate or emarginate enlarging in fruit, subcoriaceous, often much imbricate. Corolla large, red or purple, funnel-shaped with shallowly-lobed to subentire limb; the midpetaline areas often somewhat hairy and with minute glands like the leaves. Stamens inserted near the base of the corolla tube; pollen globose, spinulose. Disk annular, entire or somewhat 5-lobed. Ovary glabrous, 4-locular, each loculus with one ovule; style simple; stigma biglobular. Fruit globular, completely enclosed by the accrescent calyx; pericarp thin, breaking irregularly. Seeds 4, pubescent.

Corolla funnel-shaped, bright crimson, 4–5·5 cm. long; peduncle up to 1 cm. long
1. *beraviensis*
Corolla campanulate, pink or purple, 6·5–11 cm. long; peduncle 2·5–17 cm. long
2. *laxiflora*

1. **Stictocardia beraviensis** (Vatke) Hall. f. in Engl., Bot. Jahrb. **18**: 159 (1893).—Dammer in Engl., Pflanzenw. Ost.-Afr. **C**: 334 (1895).—Brenan, T.T.C.L.: 173 (1949).—Dandy in F. W. Andr., Fl. Pl. Anglo-Egypt. Sudan **3**: 127 (1956).—Verdc. in Kew Bull. **15**: 5 (1961); in F.T.E.A., Convolvulaceae: 69, fig. 19 (1963).—Heine in F.W.T.A. ed. 2, **2**: 352 (1963). Type from Madagascar.
   Ipomoea beraviensis Vatke in Linnaea **43**: 514 (1882). Type as above.
   Argyreia? beraviensis (Vatke) Baker in F.T.A. **4**: 201 (1906). Type as above.
   Argyrei bagshawei Rendle in Engl., Bot. Jahrb. **46**: 183 (1908). Type from Uganda.

Strong woody twiner, climbing into crowns of trees. Stems densely pubescent when young and glabrescent yellowish later. Leaf lamina ovate up to 16 × 14 cm. acute or blunt at the mucronate apex, shallowly cordate or truncate, glabrous or very sparingly pilose above, very densely grey-velvety beneath to glabrous or glabrescent; lateral nerves closely parallel; petiole up to 17 cm. long. Inflorescence axillary, many-flowered; peduncle up to 1 cm. long; pedicels up to 1 cm. long; bracts cucullate. Sepals elliptic or subcircular, up to 10 × 8 cm. long, blunt or emarginate, mucronulate, glabrescent above, hairy towards the base. Buds finely pubescent at the apices. Corolla very bright crimson with base of the tube orange-yellow, funnel-shaped, not markedly narrowed basally, 4–5·5 cm. long, with tuft of hairs at the apices of the midpetaline areas. Capsule globose, up to 2 cm. in diam., straw-coloured, woody. Seeds black, more or less grannular.

Zambia C: Mpongwe, fr. 12.ix.1965, *Fanshawe* 9320 (SRGH). W: Solwezi, between Lualaba and Mwombezhi-Lualaba, fl. 12.vi.1962, *Holmes* 1462 (K; LISC; SRGH).
From Ethiopia to West Africa and Madagascar. Riverine Forest.

2. **Stictocardia laxiflora** (Baker) Hall. f. in Bull. Herb. Boiss. **6**: 548 (1898).—Brenan, T.T.C.L.: 173 (1949).—Verdc. in Kew Bull. **15**: 5 (1961); in F.T.E.A., Convolvulaceae: 71 (1963).—Binns, H.C.L.M.: 41 (1968). TAB. **15**. Type: Malawi, Shire Highlands, *Buchanan* 388 (K, holotype).

Vigorous climber, up to 15 m. or more. Stems dark-coloured, woody, slender and pubescent. Leaf lamina ovate, up to 25 × 19 cm., acuminate, subacute and mucronate at the apex, cordate at the base, very similar to those of the *S. beraviensis*; petiole about 10 cm. long. Inflorescences axillary, few-flowered; peduncle 2·5–17 cm. long; pedicels up to 5 cm. long. Sepals broadly elliptic or subcircular, about 10 mm. long and wide blunt and emarginate, the outer ones appressed pilose outside, much imbricate. Corolla usually pink or purple, campanulate, very considerably narrowed at the base, 6·5–11 cm. long, pubescent or pilose at the apex of the midpetaline areas. Capsule not seen.

Leaf lamina glabrous to velvety, up to 25 × 19 cm; peduncle 2·5–17 cm. long; pedicels up to 5 cm. long; corolla 6·5–11 cm. long   -   -   -   -   var. *laxiflora*
Leaf lamina glabrous or sparsely pubescent, up to 14 × 12.5 cm; peduncle 0·8–5·5 cm. long; pedicels 1–2 cm. long; corolla about 7 cm. long   -   -   -   var. *woodii*

Var. laxiflora Verdec. in F.T.E.A., Convolvulaceae: 71 (1963).
   Argyreia? laxiflora Baker in Kew Bull. **1894**: 67 (1894).—Baker & Rendle in F.T.A. **4**: 200 (1906). Type as for *S. laxiflora*.
   Ipomoea buchananii Baker in Kew Bull. **1894**: 73 (1894). Type: Malawi, *Buchanan* 319 (K, holotype).
   Rivea pringsheimiana Dammer in Engl., Pflanzenw. Ost.-Afr. **C**: 334 (1895). Type from Tanzania.
   Stictocardia prinqsheimiana (Dammer) Hall. f. in Bull. Herb. Boiss. **6**: 548 (1898). Type as above.

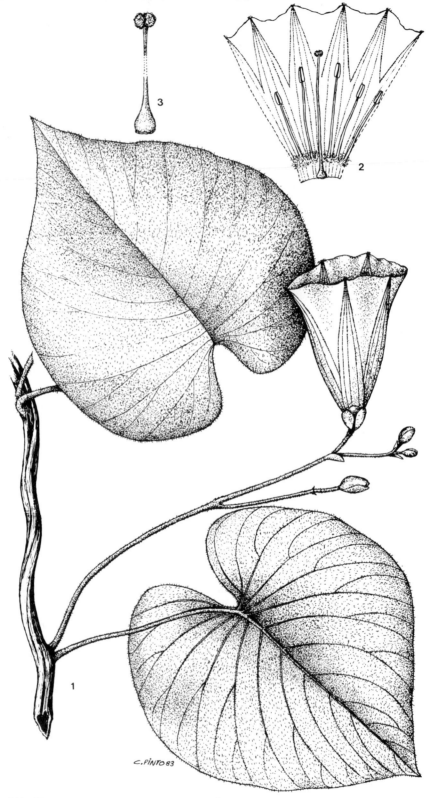

Tab. 15. STICTOCARDIA LAXIFLORA. 1, habit (×½); 2, corolla opened to show stamens and pistil (×½); 3, pistil (×1). All from *Barbosa* 1119.

Ipomoea pringsheimiana (Dammer) Rendle in F.T.A. **4**: 185 (1905); in Brenan, T.T.C.L.: 172 (1949). Type as above.

Stictocardia beraviensis (Vatke) Hall. f. subsp. *laxiflora* (Baker) Verdc. in Kew Bull. **13**: 189 (1958). Type as for species.

Zimbabwe. E: Mutare, Burma Valley 610 m., fl. 19.v.1957, *Chase* 6505 (K; SRGH). **Malawi.** S: Zomba, Mlungusi Stream, fl. 18.iv.1964, *Salubeni* 304 (K; SRGH). **Mozambique.** Z: Morrumbala Mt., fl. 15.v.1943, *Torre* 5322 (LISC). MS: Chimoio, Bandula forest, 610 m., fl. 24.iv.1958, *Chase* 6877 (BM; K; LISC; SRGH). GI: Caniçado, Guijá, Limpopo, riverside, fl. 13.xii. 1940, *Torre* 2384 (EA; K; LISC; SRGH).

Also from Tanzania. Rain and riverine forest; 600–1250 m.

Var. **woodii** (N.E. Br.) Verdc. in Kew Bull. **17**: 173 (1963); in—F.T.E.A., Convolvulaceae: 71 (1963).—Ross, Fl. Natal: 296 (1972). Type from S. Africa.

Ipomoea woodii N.E. Br. in Kew Bull. **1894**: 101 (1894). Type as above.

Stictocardia woodii (N.E. Br.) Hall. f. in Bull. Herb. Boiss. **6**: 548 (1898).—Verdc. in Kew Bull. **13**: 189 (1958).—Meeuse in Bothalia **6**: 773 (1958). Type as above.

Mozambique. MS: 16 km. N. of Sofala, fl. vii.1959, *Leach* 9194 K; SRGH). M: Maputo, fl. 26.ii.1947, *Hornby* 2585 (PRE; SRGH).

Also in Natal. Edge of riverine fringe, evergreen bush.

16. ASTRIPOMOEA Meeuse

Astripomoea Meeuse in Bothalia 6: 709 (1958).
Astrochlaena Hall. f. in Engl., Bot. Jahrb. **18**: 120 (1893) non *Astrochlaena* Corda (1845) nec Garcke (1850) *nom-illegit*.

Annual or perennial, erect or trailing herbs or subshrubs, covered with a conspicuous indumentum of soft stellate hairs on all vegetative parts, peduncles, bracteoles, pedicels and sepals. Stems usually simple or branched upwards, often firm to stout. Leaves petiolate; lamina usually ovate, oblong to subcordate, entire or coarsely dentate-sinuate. Inflorescences cymose, few to many-flowered or by reduction occasionally 1-flowered; cymes axillary but often forming terminal leafy panicels at the apices of the stems; bracteoles often small; pedicels usually short. Sepals 5, usually more or less unequal, often ovate or oblong to lanceolate; outer ones often dorsally subcarinate. Corolla funnel-shaped, purple or violet or white with a purple centre, limb spreading almost entire; midpetaline areas well-defined. Stamens included, unequal in length. Pollen spherical spinulose. Disk annular at the base of the ovary. Ovary bilocular, 4-ovuled. Stigmas bicapitate or oblong. Capsule 4-valved. Seeds 4, hairy.

1. Perennial herb; stems several to many arising separately from a woody rootstock; corolla entirely rose or mauve (rarely white) - - - - - - - 1. *malvacea*
— Annuals or short-lived perennial; stems single often considerably branched at the base; corolla whitish with a mauve centre - - - - - - - 2
2. Corolla up to 1·5 cm. long; cymes crowded, about 6-flowered - - - 2. *lachnosperma*
— Corolla over 2 cm. long; cymes lax - - - - - 3. *hyoscyamoides*

1. **Astripomoea malvacea** (Klotzsch) Meeuse in Bothalia **6**: 710 (1958).—Verdc. in Kew Bull. **13**: 192 (1958); in F.T.E.A., Convolvulaceae: 74 (1963).—Heine in F.W.T.A. ed. 2, **2**: 344 (1963).—Binns, H.C.L.M.: 39 (1968).—Ross, Fl. Natal: 295(1972).—Jacobsen in Kirkia **9**: 171 (1973). TAB. **16**. Types: Mozambique, Inhambane and Sena, *Peters* (B, syntypes†).

Breweria malvacea Klotzsch in Peters, Reise Mossamb., Bot.: 245, t. 37 (1861). Type as above.

Astrochlaena malvacea (Klotzsch) Hall. f. in Engl., Bot. Jahrb. **18**: 121 (1893).—Baker & Rendle in F.T.A. **4**: 121 (1905).—Hutch & Dalz., F.W.T.A. **2**: 213: (1931). Type as above.

An extremely variable subshrub-like perennial herb. Stems up to 2 m. long, several, erect or prostrate, more or less densely stellately tomentose, glabrescent, arising from a woody rootstock. Leaf lamina elliptic to broadly ovate, ovate-lanceolate or subrhomboid, 3–13·5 × 2–11 cm., entire or repand, acute or rounded and mucronulate at the apex, cuneate to subcordate at the base, dull green drying brownish and sparsely stellate-hairy to glabrescent above, matted with white stellate tomentum beneath; petiole varying considerably in length, up to 7·5 cm. long, densely stellate-hairy as are peduncles, pedicels and calyx. Inflorescences axillary, often forming a leafy panicle at the tops of stems; peduncle 0·7 cm. long, erect to patent, subumbellately 1 to few-flowered; pedicels up to 1·5 cm. long; bracts elliptic to ovate, up to 12 × 7 mm.,

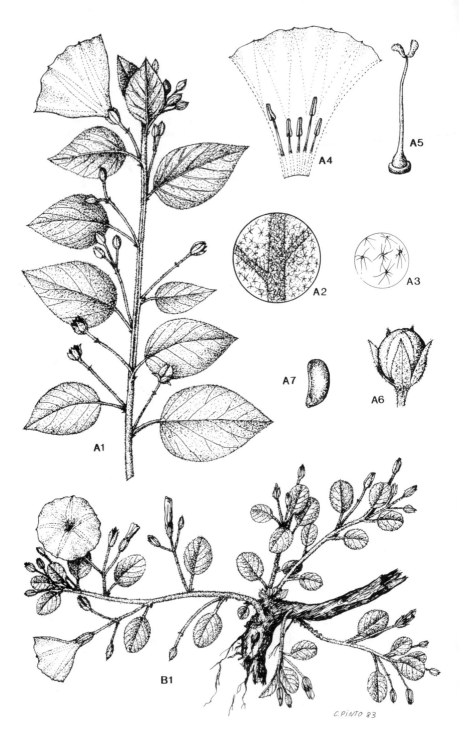

Tab. 16. ASTRIPOMOEA MALVACEA var. MALVACEA. A1, habit (×½); A2, portion of leaf showing stellate hairs (×5); A3, stellate hairs (×20); A4, corolla opened to show stamens (×1); A5, pistil (×2); A6, fruit (×2); A7, seed (×2½) A1–7 from *Barbosa & Lemos* 7955; B1, habit of A. MALVACEA var. VOLKENSII (×½), from *Brummitt* 11610.

midrib prominent. Sepals lanceolate to circular, up to 10–8 mm., acute or obtuse at the apex, tomentose outside, glabrescent inside, midrib prominent. Corolla funnel-shaped with a narrow tube, 2·5–5·5 cm., mauve or purple, glabrous or nearly so. Capsule subglobose, glabrous. Seed ovoid, compressed, blackish-brown minutely velvety-pulverulent with a tuft of fawn hairs around the hilum.

Stems often erect; leaves ovate-lanceolate, rhomboid or ovate, often more pointed at the apex but in some erect plants circular and blunt, often much exceeding 4 cm. in length   ·   ·   ·   ·   ·   ·   ·   ·   ·   ·   ·   ·   var. *malvacea*
Stems prostrate or straggling; leaves usually ovate-circular, more or less cordate at the base and obtuse or rounded at the apex, about 4 cm. or less in length   var. *volkensii*

Var. **malvacea** Verdc. in Kew Bull. **13**: 192 (1958); in F.T.E.A., Convolvulaceae: 74 (1963). —Binns, H.C.L.M.: 39 (1968).

Ipomoea floccosa Vatke in Linnaea **43**: 512 (1882). Type from Kenya.
Astrochlaena floccosa (Vatke) Hall. f. in Engl., Bot. Jahrb. **18**: 121 (1893).—Baker & Rendle in F.T.A. **4**: 123 (1905). Type as above.
Astrochlaena engleriana Dammer in Engl., Pflanzenw. Ost.-Afr. **C**: 330 (1895).—Baker & Rendle in F.T.A. **4** (2): 121 (1905). Type from Tanzania.
Astrochlaena stuhlmannii Hall. f. in Engl., Bot. Jahrb. **30**: 386 (1901).—Baker & Rendle in F.T.A. **4**: 122 (1905). Type from Tanzania.
Astrochlaena malvacea (Klotzsch) Hall. f. var. *epedunculata* Rendle in Journ. Bot. (Lond.) **39**: 59 (1901).—Eyles in Trans. Roy. Soc. S. Afr. **5**: 453 (1916). Type: Zimbabwe, Harare, *Rand 511* (BM, holotype).
Astrochlaena kaessneri Rendle in F.T.A. **4**: 123 (1905). Type from Kenya.
Astripomoea malvacea (Klotzsch) Meeuse var. *epedunculata* (Rendle) Verdc. in Kew Bull. **13**: 194 (1958); in F.T.E.A., Convolvulaceae: 76 (1963). Type as for *Astrochlaena malvacea*.
Astripomoea malvacea (Klotzsch) Meeuse var. *floccosa* (Vatke) Verdc. in Kew Bull. **13**: 195 (1958); in F.T.E.A., Convolvulaceae: 76 (1963).—Binns, H.C.L.M.: 39 (1968). Type as above.

Zambia. N: Mporokoso, near Bulaya, 975 m., fl. & fr. 13.viii.1962, *Tyrer 445* (SRGH). W: Kitwe, fl. & fr. 30.viii.1959, *Fanshawe 5198* (K). C: Lusaka, N. of Lusaka on the Great North Rd., fl. & fr. 9.viii.1952, *Angus 180* (FHQ; K). S: Kafue R., fl. & fr. 12.vii.1930, *Hutchinson & Gillett 3566* (K; LISC; SRGH). **Zimbabwe**. N: Lomagundi, 1174 m., fl. & fr. 9.x.1962, *Jacobsen 1812* (PRE). C: Makoni, Rusape, 1450 m., fl. & fr. 26.x.1930, *Fries, Norlindh & Weimarck 2295* (K; PRE; SRGH). E: Chimanimani, Haroni-Lusitu junction, fl. & fr. 12.i.1969, *Mavi 905* (K; SRGH). S: Chibi, Sikanajena Hills, 610 m., fl. & & fr. xii.1955, *Davies 1793* (K; PRE; SRGH). **Malawi**. N: Karonga, 16 km. S. of Karonga, 500 m., fl. & fr. 16.vii.1970, *Brummitt 12152* (K; LISC; MAL; SRGH). C: Lilongwe, Dzalanyama Forest Reserve, from Sinyala to Chaulongwe Falls, fl. & fr. 28.iii.1970, *Brummitt 9460* (K; SRGH). S: Ntcheu, Kirk Range Mts., fl. & fr. 12.vi.1967, *Salubeni 740* (K; SRGH). **Mozambique**. N: Mandimba, 760 m., fl. & fr. 20.ii.1942, *Hornby 4507* (K; PRE; LMA). Z: Mocuba, Namagoa, 76 m., fl. & fr. 1.iii.1943, *Faulkner 150* (K; PRE). MS: Chemba, Marínguè, fl. & fr. 12.vii.1969, *Leach & Cannell 14319* (K; LISC; SRGH). GI: Inharrime, Nhacongo, fl. & fr. 25.x.1947, *Barbosa 530* (LMA; SRGH). M: Manhiça, between Bobole and Alvor, fl. & fr. 3.x.1957, *Barbosa & Lemos 7955* (COI; K; LISC; LMA).

Also in W. E. and S. Africa. Open woodland, savanna, bushland, grassland, cultivated ground, riversides, roadsides, sandy soil and waterside swamps; 60–1750 m.

Var. **volkensii** (Dammer) Verdc., in Kew Bull. **13**: 193 (1958); in F.T.E.A., Convolvulaceae: 76 (1963). Type from Tanzania.

Astrochlaena volkensii Dammer in Engl., Pflanzenw. Ost.—Afr. **C**: 311 (1895).—Baker & Rendle in F.T.A. **4**: 120 (1905). Type as above.
Convolvulus phillipsiae Baker in Kew Bull. **1895**: 221 (1895). Type from the Somali Republic.
Astrochlaena phillipsiae (Baker) Rendle in F.T.A. **4**: 121 (1905). Type as above.
Astrochlaena stuhlmannii Hall. f. var. *parviflora* Rendle in F.T.A. **4**: 122 (1905). Type from Uganda.
Atrochlaena rotundata Pilger in Engl., Bot. Jahrb. **45**: 222 (1910). Types from Namibia.
Astrochlaena magisii De Wild. in Bull. Jard. Bot. Brux. **3**: 260 (1911). Type from Zaire.
Astrochlaena mildbraedii Pilger in Wiss. Ergebn. Zweit. Deutsch. Zentr.—Afr. Exp.: 277 (1911). Type from Tanzania.
Astrochlaena ledermannii Pilger in Engl., Bot. Jahrb. **48**: 350 (1912). Type from Cameroon.
Astrochlaena menispermoides Standl. in Smith's. Misc. Coll. **68**, 5: 11 (1917). Type from Kenya.
Astrochlaena rotundata (Pilger) Meeuse in Bothalia **6**: 711 (1958). Type as for *Astrochlaena rotundata*.

Zambia. N: Mbala, old Kasama Rd., 1500 m., fl. & fr. 21.vii.1970, *Richards* 12877 (K; SRGH). W: Kitwe, fl. & fr. 16.v.1964, *Fanshawe* 8631 (K). C: Mkushi, fl. & fr. 24.ix.1957, *Fanshawe* 3715 (K). S: Mumbwa, 30.vii.1963, *van Rensberg* 2373 (K; SRGH). **Zimbabwe**. N: Murewa, Chitowa II area, fl. & fr.x 9.iv.1966, *Mavi* 64 (SRGH). C: Gweru, Old Dog Ranch, fl. & fr.12.i.1963, *Loveridge* 547 (K; SRGH). S: Masvingo, 32 km. N. of Masvingo, fl. & fr. 4.v.1962, *Drummond* 7963 (K; SRGH). **Malawi**. C: Kasungu Distr., NW. of Kasungu Game Reserve, 1000 m., fl. & fr. 21.vi.1970, *Brummitt* 11610 (K). **Mozambique**. N: between Mocímboa da Praia and Palma, fl. & fr. 15.ix,1948, *Barbosa* 2128 (LISC; LMA). M: Maputo, fl. & fr. 25.x.1919, *Shantz* 335 (K).

Tropical Africa and S. Africa. Woodland, savanna, woodland, grassland, open bushland, cultivated ground, sandy soil and roadside; 1000–1530 m.

2. **Astripomoea lachnosperma** (Choisy) Meeuse in Bothalia **6**: 710 (1958).—Verdc. in Kew Bull. **13**: 195 (1958); in F.T.E.A., Convolvulaceae: 77 (1963).—White, F.F.N.R.: 361 (1962).—Heine in F.W.T.A. ed. 2, **2**: 344 (1963).—Amico & Bavazzano, Contr. Con. Fl. Zambezia inf. (Mozamb.) no. **6**: 27 (1968) & no. 49 (1978).—Binns, H.C.L.M.: 39 (1968). Type from the Sudan.

Ipomoea lachnosperma Choisy in DC., Prodr. **9**: 356 (1845). Type as above.

Astrochlaena lachnosperma (Choisy) Hall. f. in Engl., Bot. Jahrb. **18**: 121 (1893).—Baker & Rendle in F.T.A. **4**: 119 (1905).—N.E. Br., in Kew Bull. **1909**: 124 (1909).—Hutch. & Dalz., F.W.T.A. **2**: 213 (1931). Type as above.

Astrochlaena solanacea Hall. f. in Engl., Bot. Jahrb. **18**: 121 (1893).—Baker & Rendle in F.T.A. **4**: 120 (1905). Type from Tanzania.

Astrochlaena grantii sensu Rendle in F.T.A. **4**: 123 (1905) quoad spec. ex Chopeh leg. *Speke & Grant* non *Rendle*.

Annual with woody base. Stems simple, erect up to 120 cm. tall, covered with pale fawn to whitish stellate tomentum, as are petioles, peduncles, pedicels and calyces, becoming woody and glabrous at the base. Leaf lamina ovate or rhombic, varying to broadly elliptic, ovate-lanceolate or subrhombic 3–13 × 1·8–9 cm., acute or rounded at the apex, cuneate or truncate at the base, entire or slightly or occasionally distinctly repand above the middle, thinly covered with white stellate hairs above, more or less densely so and paler beneath; petiole up to 4 cm. long. Flowers in crowded about 6-flowered cymes; peduncle up to 5·5 cm. long; pedicels about 5 mm. long. Sepals lanceolate, ovate-lanceolate, acute, up to 6 mm. long, tomentose outside. Corolla cream to white with purple centre, funnel-shaped, up to 1·5 cm. long; limb about 2 cm. in diam. Capsule ovoid, glabrous. Seeds villous-sericeous.

Botswana. N: Maun, fl. & fr. 27.iv.1967, *Lambrecht* 155 (K; LISC; SRGH). **Zambia**. C: Katondwe, fl. & fr. 20.vi.1966, *Fanshawe* 9731 (K; SRGH). S: Kalomo, fl. & fr. 16.ii.1965, *Fanshawe* 9103 (K; SRGH). **Zimbabwe**. N: Urungwe, Zambezi Valley, 520 m., fl. & fr. 25.ii.1953, *Wild* 4078 (K; LISC; SRGH). W: Nyamandhlovu, fl. & fr. x.1953, *Orpen* 24/53 (K; SRGH). E: Chimanimani, 610 m., fl. & fr. 24.ii.1952, *Chase* 4393 (K; LISC; SRGH). **Mozambique**. T: Tete, Chicoa, fl. & fr. 26.ii.1972, *Aguiar Macedo* 4913 (COI; LISC; LMA; SRGH).

Widley spread from, Ethiopia to Nigeria, Uganda, Kenya, Tanzania and Namibia. *Mopane* woodland, open forest, grassland, roadside and sandy soil; 350–1300 m.

3. **Astripomoea hyoscyamoides** (Vatke) Verdc. in Kew Bull. **13**: 196 (1958); in F.T.E.A., Convolvulaceae: 78, fig. 20 (1963). Type from Kenya.

Shrubby short-lived perennial or annual. Stems up to 2 m. tall, often much-branched above the base, more or less hollow, tomentose or densely hairy. Leaf lamina elliptic or elliptic-oblanceolate (rarely obovate-rhombic) borne normally on the stem or high on the peduncles near base of pedicels, 2–7 × 1·5–3 cm., acute and apiculate at the apex, cuneate at the base, entire or wavy above the middle, grey-tomentose on both surfaces; petiole about 1 cm. long. Flowers in few to several-flowered umbel-like cymes or heads; peduncle 0·6–4 cm. long; pedicels 4–13 mm. long; inflorescences sometimes compound, being monochasially branched and continued beyond the first leaf and cyme; bracts small, elliptic-lanceolate or large, ovate-cordate, with prominent midrib. Sepals ovate-lanceolate, 5–10 mm. long, acute. Corolla white with purpole tube, funnel-shaped up to 3·5 cm. long. Capsule globose, yellowish. Seeds ovoid-subtrigonous, brown, pubescent.

Var. **hyoscyamoides**

Convolvulus hyoscyamoides Vatke in Linnaea **43**: 520 (1882). Type as for the species.

Astrochlaena hyoscyamoides (Vatke) Hall. f. in Engl., Bot. Jahrb. **18**: 121 (1893).—Baker & Rendle in F.T.A. **4**: 125 (1905).—Brenan, T.T.C.L.: 169 (1949). Type as above.
Astrochlaena whytei Rendle in F.T.A. **4**: 124 (1905). Type from Kenya.
Astrochlaena annua Rendle in F.T.A. **4**: 125 (1905). Type from Tanzania.

Malawi. N: Chipita, Chinunka, 1280 m., fl. & fr. 26.iv.1977, *Pawek* 12681 (SRGH).
Also from Kenya and Tanzania. Cultivated land; 900–1280 m.

The var. *melandrioides* (Hall. f.) Verdc. has stems densely hairy with spreading long-rayed stellate hairs; the bracts are larger, ovate, 1·8–2 × 1 cm., cordate acuminate and the inflorescences denser and compact. It occurs in Kenya and Tanzania.

17. IPOMOEA L.

Ipomoea L., SP. PL.: 159 (1753) & Gen. Pl., ed. 5: 76 (1754).

Annuals, biennials or perennials, herbaceous, suffruticose or sometimes woody plants. Stems woody or herbaceous, usually twining or prostrate, sometimes erect. Leaves petiolate or rarely sessile exceedingly variable in shape and size, entire or lobed to deeply divided, simple or rarely compound; pseudo-stipules sometimes present. Flowers axillary, solitary or in few to many-flowered cymes, sometimes aggregated at the apices of shoots pseudo-capitate, or enclosed in an involucre or in cymes, rarely forming a terminal leafy panicle; peduncles short or long, rarely almost nil; pedicels usually distinct, rarely almost nil; bracts and bracteoles various, sometimes leafy, free or forming an involucre. Sepals 5, very variable in size and shape, often unequal, membranaceous to coriaceous, persistent, often more or less enlarged in fruit. Corolla usually regular, rarely faintly zygomorphic, usually funnel-shaped or the tube somewhat campanulate, sometimes hypocrateriform, usually not conspicuously lobed, variously coloured. Stamens 5, usually unequal, inserted near the base of the corolla tube, subincluded or sometimes exserted; filaments filiform or somewhat linear, often dilated and hairy or papillate at the base; anthers ovate to linear; pollen globose, spinose or spinulose. Disk annular. Ovary 2-3-4, or rarely 5-locular, usually 4-ovuled, rarely with 6–10 ovules; style filiform, included to exserted; stigma 2-globular. Capsule globose to avoid, 3–10-valved. Seeds usually 4 (rarely less by abortion or up to 10), glabrous, pubescent or clothed (sometimes only partly) with very long hairs.

A very large cosmopolitan genus of at least 500 species, mainly restricted to the tropics. The following subgenera and sections are represented in this account.

ORTHIPOMOEA Choisy (syn. *Calycanthemum* (Klotzsch) Hall. f.	- -	spp. 1–17
DASYCHAETIA Hall. f.	- -	spp. 18–20
IPOMOEA		
Involucrata Baker & Rendle	- -	spp. 21–22
Ipomoea (syn. *Cephalanthae* (Choisy) Hall. f.)	- -	spp. 23–32
Pharbitis (Choisy) Griseb. (syn. *Chorisanthae* Hall. f.)	- -	spp. 33–37
BATATAS (Choisy) Griseb.	- -	sp. 38
QUAMOCLIT (Moench) Hall. f. (Syn. *Leiocalyx* Hall. f. sensu lato)		
Erpipomoea Choisy (Syn. *Leiocalyx* Hall. f. sensu stricto)	- -	spp. 39–66
Calonyction (Choisy) Hall. f.	- -	spp. 67–68
Quamoclit	- -	sp. 69
ERIOSPERMUM Hall. f.		
Eriospermum	- -	spp. 70–74
Acmostemom (Pilger) Verdc.	- -	spp. 75–78
POLIOTHAMNUS (Hall. f.) Verdc. (including *Argyrophyllae* Baker & Rendle & *Floriferae* Baker & Rendle)	- -	spp. 79–81

Species escaped from cultivation and well naturalised have been included in the main part of the text. Some of the more important purely ornamental species and the respective key are mentioned here.

Key to ornamental cultivated species

1. Woody plants 1–9·5 m. tall   ·   ·   ·   ·   ·   ·   ·   ·   ·   **2**
— Twining plants, often annuals   ·   ·   ·   ·   ·   ·   ·   ·   **4**
2. Plants glabrous; sepals about 10 mm. long; corolla white, glandular-hairy within tube, up to 9·5 cm. long   ·   ·   ·   ·   ·   ·   ·   ·   *intrapilosa*
— Plant minutely to densely pubescent sepals 5–13 mm. long; corolla pink or white, not glandular-hairy ·   ·   ·   ·   ·   ·   ·   ·   ·   **3**

3. Sepals round, 5–6 mm. long, minutely pubescent; leaves long and pointed, minutely pubescent; corolla pink, up to 9 cm. long - - - - - *carnea* subsp. *fistulosa*
— Sepals ovate, 13 mm. long, velutinous; leaves acute, velvety-pubescent; corolla white, about 5 cm. long - - - - - - - - - - - *arborescens*
4. Leaves ovate, cordate at the base, long acuminate; not divided or deeply lobed corolla violet-blue or purple, with a white tube, up to 7 cm. long - - - - *tricolor*
— Leaves pinnately parted or deeply palmately lobed - - - - 5
5. Leaves pinnately parted into numerous linear or filiform segments; corolla red-crimson (white in var. *albiflora* G. Don), about 3 cm. long - - - - - - *quamoclit*
— Leaves deeply palmately lobed to beyond the middle or to the base into 3–5 segments; corolla red or magenta, 4 cm. long - - - - - - - - - *horsfalliae*

Ipomoea arborescens (H.B.K.) G. Don, an erect, woody, tree-like plant, reaching 6 m. in height, native of Mexico, is cultivated in Zimbabwe.

Ipomoea carnea Jacq. subsp. **fistulosa** (Mart. ex Choisy) D. Austin, is a subshrubby plant up to 3 m. high, a native of Brazil and cultivated in Zambia, Zimbabwe, Malawi and Mozambique, some specimens seen are probably escapes. *Ipomoea crassicaulis* (Benth.) Robinson and *Ipomoea fistulosa* Mart. ex Choisy are synonymous.

Ipomoea horsfalliae Hook., a large glabrous twiner with the adult stems woody, is native of the West Indian Islands and cultivated in Zimbabwe.

Ipomoea intrapilosa Rose, possibly not distinct from *I. wolcottiana* Rose, is a shrub or tree up to 9·5 m. in height, native of Mexico and cultivated in Zambia, Zimbabwe and Mozambique.

Ipomoea quamoclit L., an annual glabrous twiner, rarely prostrate, is native to India and cultivated in Malawi and Mozambique.

Ipomoea tricolor Cav., a herbaceous glabrous twiner a native of tropical America, is not uncommon as an escape in Zambia, Zimbabwe, Malawi and Mozambique.

Key to native and naturalised species

Some characters used, e.g. lobing of the leaves, are variable in a single species; in such instances the species has been inserted twice.

1. Sepals very distinctly awned - - - - - - - - - - - 2
— Sepals not awned but often very long-acuminate or mucronate - - - - 4
2. Flowers scarlet - - - - - - - - - - - 69. *hederifolia*
— Flowers white or bluish-purple - - - - - - - - - 3
3. Stem usually smooth; corolla white or greenish-cream, tube 7–12 cm. long - 67. *alba*
— Stem covered with laterally flattened expansions; corolla bluish purple, tube 3–6 cm. long
   68. *turbinata*
4. Leaf lamina spathulate with circular apical portion and narrowly obtriangular lower portion; sepals conspicuously 3-nerved - - - - - - - 65. *trinervia*
— Leaf lamina and sepals not as above - - - - - - - - 5
5. Indumentum stellate-pubescent; corolla up to 3 cm. long, tube purple, limb whitish
   13. *ephemera*
— Indumentum not as above - - - - - - - - - - - 6
6. Flowers in heads subtended by a boat-shaped involucre - - - - - 7
— Flowers if in heads then not subtended by a boat-shaped involucre - - - 8
7. Corolla tube broadly cylindrical at base; sepals acute - - - 21. *involucrata*
— Corolla tube very narrow at base; sepals generally more or less obtuse - 22. *pileata*
8. Leaves subcoriaceous or fleshy, mostly obtuse to broadly rounded, emarginate or 2-lobed at the apex; corolla 3–6·5 cm. long - - - - - - - - - - 9
— Leaves and corolla not as above - - - - - - - - - - 11
9. Corolla white or yellowish with a purple centre; leaf lamina small, up to 5 × 3 cm., fleshy, very variable in shape often on the same plant, lanceolate, ovate or oblong, obtuse or emarginate to lobed at the apex, entire or 3–5 lobed - - - - 50. *stolonifera*
— Corolla pink or purple; leaf lamina larger, up to 9·5 × 10·5 cm., subcoriaceous, circular kidney shaped, quadrangular or elliptic, rounded or emarginate to 2-lobed at the apex, entire - - - - - - - - - - - - - - - 10
10. Leaf lamina reniform, rounded at the apex, base cordate, sepals unequal; seeds glabrous
    51. *asarifolia*
— Leaf lamina subcircular elliptic, obreniform or quadrangular, emarginate at the apex, usually appearing deeply 2-lobed, base truncate, cuneate or subcordate; sepals subequal; seeds tomentose-villous - - - - - - - - - 52. *pes-caprae*
11. Leaves compound, divided into 5–7 narrowly lanceolate leaflets 7–8·5 × 1–2·4 cm.
    59. *venosa* subsp. *venosa*
— Leaves entire or much divided but never compound - - - - - 12
12. Plant erect, suberect or decumbent, or if prostrate then a cultivated plant grown for its edible tubers - - - - - - - - - - - - - 13

— Plant with prostrate or twining stems; not cultivated as a food plant · · · · 30
13. Shrubs or subshrubs, more or less woody, with stout stems; inflorescences 1-3-flowered; seeds covered with long hairs · · · · · · · · · 14
— Annual or perennial herbs including those which flower as first year rosette plants or when young and suberect but later send out short trailing or ascending shoots · · · 16
14. Corolla very long, salver-shaped, white or pink; tube narrowly cylindric, 7–9 cm. long
79. *adenoides*
— Corolla funnel-shaped, white, mauve or rose-purple; tube broadly cylindric, 3.5–8 cm. long · · · · · · · · · · · · · · 15
15. Leaf lamina oblong or ovate-circular, cordate or truncate at the base; inflorescence 1–3 flowered · · · · · · · · · · · 76. *verbascoidea*
— Leaf lamina oblong to ovate, rounded or cuneate at the base; inflorescences 1-flowered
78. *prismatosyphon*
16. Leaves all lobed (or indented) or lobed (or indented) and entire on the same plant · 17
— Leaves all entire · · · · · · · · · · · · · 20
17. Annual or biennial; leaf lamina ovate, oblong or elliptic, coarsely lobed; corolla white or pink, 1·4–1·6 cm. long · · · · · · · · · · 7. *polymorpha*
— Perennial with tuberous roots · · · · · · · · · · 18
18. Seeds covered with long shiny fawn hairs; corolla 4–7·5 cm. long · · 56. *bolusiana*
— Seeds glabros or appressed-pubescent; corolla 2·3–5 cm. long· · · · · 19
19. Pilose herb; corolla funnel-shaped, 2·3–5 cm. long; seeds appressed-pubescent
9. *oenotherae*
— Glabrous herb cultivated for edible tubers; corolla campanulate, 3–4·5 cm. long; seeds glabrous · · · · · · · · · · · 38. *batatas*
20. Sepals (outer ones) verrucose-echinate; annual herb, glabrous · · 45. *humidicola*
— Sepals not as above; annuals or perennials · · · · · · · 21
21. Corolla up to 15 mm. long, white, cream or pale yellow green · · · · 22
— Corolla much longer than 15 mm. · · · · · · · · · 23
22. Leaf-lamina thinly hairy on both surfaces; capsule pilose · · 2. *leucanthemum*
— Leaf lamina appressed-pilose above, densely silvery-pilose beneath; capsule glabrous
24. *chloroneura*
23. Leaf lamina with the base cordate or cordate to truncate · · · · · 24
— Leaf lamina with the base not as above · · · · · · · · 25
24. Bracts ovate, lanceolate or linear, up to 19 mm. long; leaves pubescent above and densely so beneath; perennial herb with woody rootstock; flowers solitary or in dense cymes
14. *fulvicaulis*
— Bracts minute, lanceolate, more or less 3 mm. long; leaves pubescent or hairy on both surfaces; perennial herb without rootstock; flowers solitary or in lax cymes
17. *tenuirostris*
25. Inflorescences bracteate terminal heads; leaves with dense silvery silky appressed hairs beneath · · · · · · · · · 23. *crepidiformis*
— Inflorescences not as above · · · · · · · · · · 26
26. Sepals lanceolate or linear-lanceolate, 10–18 mm. long · · · · · 27
— Sepals ovate or ovate-lanceolate, 4·5–8 mm. long · · · · · · 29
27. Corolla funnel-shaped, 5–9·5 cm. long· · · · · · · 46. *welwitschii*
— Corolla tubular-funnel shaped, 1·8–3·7 cm. long · · · · · · 28
28. Leaf lamina or oblanceolate-obovate; corolla pink to whitish with purple centre, 2–3·5 cm. long · · · · · · · · · · · 18. *linosepala*
— Leaf lamina linear to oblong; corolla yellow or white, 1·8–3·7 cm. long · 19. *alpina*
29. Leaf lamina linear-lanceolate or oblong, 2–5 × 0·2–1 cm.; inflorescences axillary, 1-flowered · · · · · · · · · · · · 11. *recta*
— Leaf lamina from linear-oblong or lanceolate to largely lanceolate or oblong-lanceolate, 3–12·5 × 0·3–3 cm.; inflorescences axillary, 1–4 flowered · · · 48. *richardsiae*
30. Leaf lamina linear, with extremely close parallel nerves · · · · · 31
— Leaf lamina with normally spaced nerves · · · · · · · 33
31. Corolla up to 2·6 cm. long; plant more or less pubescent · · · 12. *milnei*
— Corolla 4–9·5 cm. long; plant glabrous· · · · · · · · 32
32. Inflorescences 1-flowered; corolla 4–7·5 cm. long; seeds covered with long shiny fawn hairs
56. *bolusiana*
— Inflorescences 1–3 flowered; corolla 5–9·5 cm. long; seeds shortly puberulous with a basal tuft of hairs near the hilum · · · · · · · · 46. *welwitschii*
33. Shrubs with a white somewhat floccose tomentum; flowers 6·5–14·5 cm. long; seeds with long hairs · · · · · · · · · · · · 34
— Not as above · · · · · · · · · · · · · 36
34. Leaf lamina glabrescent beneath, becoming reticulate due to a dense tomentum generally on the main nerves and veins; sepals broadly oblong, 1·5 cm. long, glabrescent but generally retaining the pubescence at the base · · · · · · 75. *albivenia*
— Leaf lamina persistently tomentose beneath; sepals elliptic, 1–2 cm. long, persistently tomentose · · · · · · · · · · · · 35
35. Bracts linear-oblong to linear-oblanceolate, 1·3–2 cm. long; leaf lamina oblong or ovate-

circular 5–18·5 × 5–17·5 cm., cordate or truncate at the base, margin entire or slightly sinuate　-　　-　　-　　-　　-　　-　　-　　-　　-　　-　　-　　76. *verbascoidea*

— Bracts obovate up to 2·5 cm. long; leaf lamina ovate-oblong, elliptic-oblong or lanceolate-oblong, 5–20·5 × 3–13 cm., broadly rounded or subtruncate at the base, margin entire, repand to deeply dentate　-　　-　　-　　-　　-　　-　　-　　-　　77. *bakeri*

36. Leaf lamina with a sparse or dense matted white indumentum beneath　-　　-　　-　　37

— Leaf lamina hairy, tomentose or glabrous, but without a matted white indumentum beneath　-　　-　　-　　-　　-　　-　　-　　-　　-　　-　　-　　-　　-　　40

37. Sepals with mixed indumentum of hairs, yellow bristles and sessile glands, but one sort sometimes absent; seeds pubescent　-　　-　　-　　-　　-　　-　　-　　-　　38

— Sepals hairy, without mixed indumentum; corolla 1·2–3·2 cm. long; seeds pubescent, hairs often in small tufts and sometimes long hairs along the angles　-　　-　　33. *magnusiana*

38. Annual; corolla 1·4–2 cm. long　-　　-　　-　　-　　-　　-　　-　　31. *dichroa*

— Perennial; corolla 2–5 cm. long　-　　-　　-　　-　　-　　-　　-　　-　　-　　39

39. Flowers generally in small to large dense heads, very rarely solitary; seeds very shortly pubescent or glabrous　-　　-　　-　　-　　-　　-　　-　　-　　-　　28. *wightii*

— Flowers generally in lax cymes, very rarely solitary; seeds yellow-pubescent, also with 2 tufts of long white hairs　-　　-　　-　　-　　-　　-　　-　　-　　-　　30. *ficifolia*

40. Sepals pilose and with sessile glands　-　　-　　-　　-　　-　　32. *pharbitiformis*

— Sepals without such a mixed indumentum　-　　-　　-　　-　　-　　-　　41

41. Sepals ovate to lanceolate, widened to the cordate or subhastate base, with a narrow lanceolate 2–10 × 1·5–10 mm. acute apex; corolla 1·3–2·2 cm. long; capsule glabrous; seeds velvety pubescent with hairs up to 5 mm. long　-　　-　　-　　-　　16. *sinensis*

— Sepals without cordate or subhastate base or, if so, then without the above characters combined　-　　-　　-　　-　　-　　-　　-　　-　　-　　-　　-　　42

42. Corolla small, 0·6–1·7 cm. long (if corolla bright yellow see 40. *I. obscura*)　-　　-　　43

— Corolla usually over 2 cm. long　-　　-　　-　　-　　-　　-　　-　　-　　-　　50

43. Leaf lamina lobed　-　　-　　-　　-　　-　　-　　-　　-　　-　　-　　44

— Leaf lamina entire　-　　-　　-　　-　　-　　-　　-　　-　　-　　-　　-　　47

44. Sepals leaf-like, ovate with narrow apical portion or vaguely 3-lobed, 6–11 mm. long; leaves mostly pinnately lobed　-　　-　　-　　-　　-　　-　　-　　-　　-　　-　　45

— Sepals membranous or thinly coriaceous, ovate-elliptic, entire, 3–5 mm. long; leaves palmately divided　-　　-　　-　　-　　-　　-　　-　　-　　-　　-　　-　　46

45. Corolla very narrowly tubular, more or less 10 mm. long; capsule 6-valved; seeds up to 6　　　　　　　　　　　　　　　　　　　　　　　　8. *kotschyana*

— Corolla narrowly funnel-shaped, 14–16 mm. long; capsule 4-valved; seeds 4 or fewer　　　　　　　　　　　　　　　　　　　　　　　　7. *polymorpha*

46. Lobes of leaves entire; capsule 4-valved; seeds with velvety pubescence and 5–10 mm. long hairs　-　　-　　-　　-　　-　　-　　-　　-　　-　　-　　-　　62. *tenuipes*

— Lobes of leaves acutely serrate; capsule 6-valved; seeds tomentose　-　　-　　63. *coptica*

47. Leaf lamina ovate, hastate or cordate at the base　-　　-　　-　　-　　-　　48

— Leaf lamina elliptic, rounded or only slightly cordate at the base　-　　-　　-　　49

48. Sepals hairy; capsule hairy; seeds finely punctate, glabrous　-　　-　　1. *eriocarpa*

— Sepals glabrous or pubescent, capsule glabrous; seeds shortly pubescent　-　　5. *plebeia*

49. Leaves with silvery appressed hairs beneath, very discolorous; corolla pale yellow green or white　-　　-　　-　　-　　-　　-　　-　　-　　-　　-　　24. *chloroneura*

— Leaves glabrous or pubescent, not discolorous; corolla red or white　　　　　　　　　　　　　　　　　　　　　　　　14. *coscinosperma*

50. Leaf laminas all lobed or lobed and entire on the same plant　-　　-　　-　　-　　51

— Leaf laminas all entire or margins crenate to crenate-dentate or rarely sinuate below　　64

51. Leaf lamina 5-lobed or coarsely 5-toothed; sepals very conspicuously spathulate, 3-nerved　　　　　　　　　　　　　　　　　　　　　　　66. *simonsiana*

— Leaf lamina and sepals without the above characters combined　-　　-　　-　　52

52. Leaf lamina all palmately or shallowly lobed and entire on the same plant　-　　-　　53

— Leaf lamina palmately deeply 5–9 lobed; lobes narrow, 0·1–3·5 cm. wide　-　　-　　59

53. Leaf lobes broad, 1·7–4·2 cm. wide and if narrow then only 3 lobes　-　　-　　-　　54

— Leaf lobes narrow, less than 1 cm. wide and more than 3 lobes　-　　-　　-　　-　　58

54. Plant a glabrous liane; calyx ovoid; sepals circular or elliptic, strongly, convex, tightly closed about the narrow base of the corolla tube　-　　-　　-　　-　　-　　73. *mauritiana*

— Plant never entirely glabrous; sepals spreading, not tightly gripping the corolla, ovate or linear-lanceolate　-　　-　　-　　-　　-　　-　　-　　-　　-　　-　　-　　55

55. Sepals ovate more or less obtuse; leaves usually entire, only rarely lobed; seeds with long hairs　-　　-　　-　　-　　-　　-　　-　　-　　-　　-　　-　　72. *rubens*

— Sepals lanceolate or elliptic with attenuated linear apices; seeds glabrescent or pubescent 56

56. Sepals lanceolate, narrowed above the basal part which alone is bristly　-　　-　　-　　57

— Sepals uniformly lanceolate, softly pubescent or glabrescent　-　　-　　-　　34. *indica*

57. Sepals 2·3–2·8 cm. long, the apical portions conspicuously narrow; corolla 5·5–7 cm. long　　　　　　　　　　　　　　　　　　　　　　　35. *nil*

— Sepals 1–1·5 cm. long, the apical portion not conspicuously narrow; corolla 4·5–6 cm. long　　　　　　　　　　　　　　　　　　　　　　　36. *purpurea*

58. Leaf lamina broadly cordate at the base; corolla 2–3·5 cm. long; seeds shortly pubescent
    43. *papilio*
— Leaf lamina cuneate or rotundate at the base; corolla 6 cm. long; seeds with fawn hairs more or less 10 mm. long · · · · · · · · · · · 57. *fanshawei*
59. Leaf lobes obovate, conspicuously serrate; corolla tube narrowly cylindric 1·7–2·1 cm. long
    64. *ticcopa*
— Leaf lobes elliptic or linear, entire or undulate; corolla tube not narrowly cylindric 60
60. Corolla under 2·5 cm. long; seeds densely tomentose usually with long hairs on the margins
    61. *hochstetteri*
— Corolla 3–10 cm. long · · · · · · · · · · · · 61
61. Sepals 1–2 tuberculate (spurred) at the base, smooth or verruculose; corolla yellow or white, with purple centre, 4–10 cm. long · · · · · · · 58. *tuberculata*
— Sepals quite smooth at the base; corolla not at all yellow · · · · · · 62
62. Plant with spreading bristles on stems; leaves pilose to strigose; bracts foliaceous, surrounding capitate inflorescences · · · · · · · · · 29. *pes-tigridis*
— Plant almost glabrous; stems smooth or muriculate; bracts minute; flowers solitary or in few flowered cymes; sepals ovate · · · · · · · · · · 63
63. Leaf lobes linear-lanceolate or obovate-elliptic; pseudo-stipules absent; sepals 6–7·5 mm. long; stems smooth · · · · · · · · · · · 59. *venosa*
— Leaf lobes lanceolate to ovate; pseudo-stipules present, foliaceous; sepals 4–6·5 mm. long, sometimes verruculose; stems smooth or muriculate · · · · 60. *cairica*
64. Leaf margin more or less distinctly crenate or crenate-dentate; outer sepals ovate with cordate base; corolla 1–1·3 cm. long; capsule densely pilose; seeds hairy; annual
    10. *hackeliana*
— Not as above · · · · · · · · · · · · · 65
65. Glabrous littoral plant; leaf lamina circular to ovate; corolla white or greenish-yellow, narrowly funnel-shaped, 8–10 cm. long, open at night · · · · 74. *violacea*
— Not littoral plants and without the outer characters combined · · · · · 66
66. Glabrous swamp plant with thick semi-succulent stems trailing in the water 49. *aquatica*
— Not as above · · · · · · · · · · · · · 67
67. Shrubby climbers or woody-stemmed lianas · · · · · · · · 68
— Annual or perennial twiners; seeds glabrous or shortly hairy · · · · · 74
68. Calyx of tough convex sepals tightly clasping the corolla tube; glabrous liane
    73. *mauritiana*
— Calyx not tightly clasping the corolla tube sepals not convex · · · · · 69
69. Inflorescences borne on the short older shoots, usually when the plant is leafless · 70
— Inflorescences borne with the leaves · · · · · · · · · 71
70. Leaf lamina narrowly to largely oblong, ovate or obovate, 3–7·5 × 0·5–2·5 cm., acute or rounded; sepals more or less largely ovate; shrubby with woody rootstock and trailing habit
    55. *vernalis*
— Leaf lamina ovate or circular or narrowly triangular, 3·5–8·5 × 1·3–5·5 cm., abruptly acute or acuminate; sepals oblong-lanceolate; tall woody twiner · · · 71. *shirambensis*
71. Plants glabrous; sepals ovate elliptic, 13–17 mm. long; seeds glabrous or puberulous
    70. *shupangensis*
— Plant usually villous or hairy, sometimes glabrescent; sepals ovate or linear-lanceolate, 6–25 mm long; seeds hairy (hairs sometimes 10 mm.long) · · · · 72
72. Sepals ovate, 6–8 mm. long; corolla 4–5 cm. long, sparsely pilose; seeds with hairs 2·5 mm. long · · · · · · · · · · · · · 72. *rubens*
— Sepals linear-lanceolate to ovate, 7–25 mm. long; corolla 5–8 cm.; seeds with long hairs
    73
73. Indumentum of very fine, silvery appressed hairs at least on the bracts and sepals; corolla 5–8 cm. long, white, cream or yellow with purple centre, · · · 80. *kituiensis*
— Indumentum villous, glabrescent; corolla pink, up to 6·5 cm. long · · 81. *consimilis*
74. Leaf lamina ovate, very discolorous, densely appressed silver-pilose beneath; flowers in bracteate heads; corolla up to 2·8 cm. long; seeds puberulous · · · 27. *heterotricha*
— Leaf lamina never markedly discolorous, or, if so, then without the above characters combined · · · · · · · · · · · · · 75
75. Bracts paired, oblong to ovate, situated on peduncles just or well below the inflorescence
    76
— Bracts minute or linear or scattered on the inflorescence, not large and paired · · 77
76. Indumentum typically short and dense; leaf lamina cordate to truncate at the base, pubescent above and densely so beneath; corolla mauve or purple, 2·5–5 cm. long; seeds pubescent · · · · 14. *fulvicaulis* and intermediates with var. *asperifolia*
— Indumentum typically much longer (rarely plant glabrescent); leaf lamina truncate, rounded, slightly hastate, emarginate or subcordate at the base, with long hairs on both surfaces, rarely glabrous; corolla white, red or white with purple centre, 3–6·5 cm. long; seeds glabrous or nearly so · · · · · · · · · · 15. *crassipes*
77. Leaf lamina linear or oblong to ovate, cuneate, rounded, or only shallowly cordate or hastate at the base · · · · · · · · · · · · 78
— Leaf lamina ovate to triangular, cordate, sagittate or hastate · · · · · 87

78. Flowers in pedunculate few-to many-flowered heads (rarely reduced to a single flower); sepals 1·8–3·5 cm. long, lanceolate to ovate; leaf lamina up to 14 × 7·5 cm., hairy on both surfaces; perennials with fusiform tuberous root, usually densely pubescent at least in the young parts - - - - - - - - - - - - - 79
— Not as above - - - - - - - - - - - - - 80
79. Leaf lamina ovate-lanceolate or oblong-lanceolate, gradually narrowed from a broad base towards the subobtuse to accuminate apex; corolla bright magenta, densely silky on the midpetaline areas - - - - - - - - - - 25. *ommaneyi*
— Leaf lamina from a broad base oblong to ovate elliptic, not often gradually tapering to the emarginate or rounded apex; corolla magenta, silky-strigose - - - 26. *atherstonei*
80. Corolla 1·2–1·5 cm. long, mauve or white; capsule hirsute; seeds with a villous tomentum of appressed stiff hairs; procumbent or trailing hairy herb - - - - 6. *gracilisepala*
— Corolla over 2 cm. long; capsule glabrous; seeds from glabrous to hairy; herbs of various habits - - - - - - - - - - - - - 81
81. Sepals linear to lanceolate (rarely ovate); often hairy herbs - - - - - 82
— Sepals ovate to oblong 6–8 × 2–4 mm.; usually glabrous herbs - - - - 86
82. Sepals 6·5–8 mm. long - - - - - - - - - - - 83
— Sepals 25–95 mm. long - - - - - - - - - - - 84
83. Sepals ovate-lanceolate up to 8 × 3 mm.; corolla white or pink with purple centre, up to 4·5 cm. long - - - - - - - - - - - - 11. *recta*
— Sepals lanceolate, 6·5–7 mm. long; corolla white, more or less 2·7 cm. long 20. *polhillii*
84. Herb with short stiff branchlets, glabrous or puberulous with minute hairs; often flowering when leafless; corolla 5–9·5 cm. long; seeds shortly puberulous with a basal tuft of hairs near the hilum - - - - - - - - - - - 46. *welwitschii*
— Trailing herbs with slender hairy stems and leaves, hairs sometimes restricted to the veins; corolla very narrow at base of tube, 2·5–6 cm. long - - - - - - 85
85. Sepals lanceolate 18 × 4 mm., leaves compound or minutely subcordate; flowers subsessile or peduncle up to 1·2 cm. long - - - - - - 13. *blepharophylla*
— Sepals with wider basal part 6–9 mm. long and linear apex 1–1·2 cm. long; leaves slightly subcordate; flowers pedunculate, peduncle 1–7·5 cm. long - - - 14. *fulvicaulis*
86. Flowers solitary (rarely 2 flowers); pedicle up to 2 cm. long; sepals verrucose (rarely almost smooth) - - - - - - - - - - - 47. *barteri*
— Flowers in dense cymes; peduncle 3·5–27 cm. longl sepals smooth - - 44. *lapathifolia*
87. Sepals papillose or verruculous, with or without hairs - - - - - 88
— Sepals not papillose nor verruculous - - - - - - - - 90
88. Sepals 10–18 mm. long, ovate, shortly acute, glabrous; corolla up to 6 cm. long, mauve 54. *protea*
— Sepals 6–8·5 mm. long, ovate, ovate-elliptic or circular, glabrous or hairy; corolla 1·8–7·5 cm. long - - - - - - - - - - - - 89
89. Sepals verruculous-tuberculate, ovate-elliptic, acute or obtuse, glabrous; corolla 1·8–3 cm. long, white - - - - - - - - - - 41. *verrucisepala*
— Sepals papillose, rarely almost smooth, ovate or circular obtuse, hairy or glabrescent; corolla 5–7·5 cm. long, crimson, rose or mauve or limb white and tube coloured 47. *barteri* var. *cordifolia*
90. Flowers in dense, few-to several flowered inflorescences; peduncle rather long and pedicels short - - - - - - - - - - - - - 91
— Flowers solitary or if in several-flowered inflorescences then pedicels long (1–3·5 cm.) 94
91. Corolla funnel-shaped, blue or blue-purple, 3·5–8 cm. long; naturalised herbaceous climbers - - - - - - - - - - - - 92
— Corolla funnel-shaped or almost salver-shaped, magenta, pink or almost white, 2–7·5 cm. long; wild plants - - - - - - - - - - - 93
92. Sepals lanceolate, 1·5–2·3 cm., long, caudate-acuminate; corolla 5–8 cm. long 34. *indica*
— Sepals lanceolate-oblong, c. 0·5 cm. long, mucronate; corolla 3·5–4·5 cm. long 37. *parasitica*
93. Corolla with narrow tube and almost salver-shaped, lilac-pink or almost white, 2–6 cm. long - - - - - - - - - - - - 42. *sepiaria*
— Corolla funnel-shaped, magenta, 4–7·5 cm. long - - - - - 26. *atherstonei*
94. Buds more or less glabrous except for a small apical tuft of short white hairs; leaves densely pilose - - - - - - - - - - - 53. *transvaalensis*
— Buds glabrous or pubescent, without apical tufts of hairs - - - - 95
95. Sepals lanceolate to linear or ovate to lanceolate, foliaceous; corolla white and purple, or magenta, or mauve - - - - - - - - - - - 96
— Sepals ovate to ovate-lanceolate, membranous; corolla yellow or whitish with crimson-purple eye - - - - - - - - - - - - 98
96. Corolla 1·2–1·5 cm. long, mauve or white; sepals lanceolate to linear 0·7–1·5 × 0·1–0·2 cm. - - - - - - - - - - - 6. *gracilisepala*
— Corolla 2–6 cm. long, white and purple or magenta; sepals lanceolate to ovate-lanceolate 97

97. Corolla 2–4 cm. long; sepals 0·7–1·2 cm. long, pubescent; seeds velvety sometimes with
     long hairs on the upper angles; perennial herb · · · · · 17. *tenuirostris*
—   Corolla 4·5–6 cm. long; sepals 1·0–1·5 cm. long, bristly on basal parts; seeds glabrous or
     very shortly pubescent; annual herb, ornamental · · · · · 36. *purpurea*
98. Corolla 2·7–5·5 cm. long · · · · · · · · · 39. *ochracea*
—   Corolla 1·5–2·5 cm. long · · · · · · · · · 40. *obscura*

1. **Ipomoea eriocarpa** R. Br., Prodr. Fl. Nov. Holl.,: 484 (1810).—Choisy in DC., Prodr. **9**:
     369 (1845).—Hiern, Cat. Afr. Pl. Welw. **1**: 732 (1898).—Baker & Rendle in F.T.A. **4**: 136
     (1905).—Eyles in Trans. Roy. Soc. S. Afr. **5**: 454 (1916).—van Ooststr. in Fl. Males. Ser. 1,
     4: 462, fig. 35 (1953).—Wild, Guide to Fl. Victoria Falls: 155 (1953).—Dandy in F.W.
     Andr., Fl. Anglo-Egypt, Sudan **3**: 113 (1956).—Meeuse in Bothalia **6**: 722 (1958).—Heine
     in F.W.T.A. ed. 2, **2**: 350 (1963).—Verdc. in F.T.E.A., Convolvulaceae: 91, fig. 22, 5 and
     24, 4 (1963).—Binns H.C.L.M.: 40 (1968).—Williamson, Useful Pl. Malawi: 70(1972).
     —Jacobsen in Kirkia **9**: 171 (1973). Type from Australia.
          Convolvulus hispidus Vahl, Symb. Bot. **3**: 29 (1794). Type from "India orientalis".
          Ipomoea hispida (Vahl) Roem. & Schult., Syst. Veg., ed. nov. **4**: 238 (1819) nom illegit
     non Zuccagni (1806).—Hall. f. in Engl., Bot. Jahrb. **18**: 123 (1893).—Hutch. & Dalz.,
     F.W.T.A. **2**: 216 (1931).
          Ipomoea sessiliflora Roth., Nov. Pl. Sp.: 116 (1821). Type from "India orientalis".
          Ipomoea carsonii Baker Type: Zambia, Plateau above Lake Tanganyika, *Carson* s.n. (K);
     This is possibly a hybrid between *I. eriocarpa* and *I. tenuirostris*.

Very variable annual. Stems twining or prostrate, pubescent or hispid, covered with
both long and short hairs. Leaf lamina ovate-cordate to linear-oblong, 2–10 × 0·8–7
cm., usually subhastate at the base with rounded lobes, apex long-attenuate to
acuminate, obtuse or acute, margin entire pilose-strigose or glabrescent; petiole 1–6
cm. long. Inflorescences axillary, (1)3-many flowered, subsessile or peduncle up to 1·5
cm. long; bracts linear or lanceolate, pilose; peduncle up to 1·5 cm. long; bracts linear
or lanceolate, pilose; pedicels more or less 5 mm. long, pilose. Sepals subequal, ovate,
acuminate, up to 9 mm. long, hispid-pilose, spreading in young fruit; basal part 5 × 4
mm.; apical part. 4 × 0·5 mm. Corolla tubular to funnel-shaped, mauve, white, pink or
white with a mauve centre, up to 10 mm. long, midpetaline areas pilose. Ovary with
long hairs. Capsule ovoid-globose to globose, hairy, apiculate by the style-base. Seeds
black, finely punctate, glabrous.

Botswana. N: N. of Lake Xau, fl. & fr. 22.iii.1965, *Wild & Drumond* 7229, (K; LISC;
SRGH). **Zambia**. N: Mbala, Sansia Falls, Kalambo, 1530 m., fl. & fr. 7.v.1969, *Sanane* 683
(K). W: Mwinilunga Distr., slope E. of Matonchi Farm, fl. & fr. 11.xi.1937, *Milne-Redhead* 3197
(BM; K). C: Lusaka Distr., Chinyunyu Hot springs, 85 km. from Lusaka on Rd. to Petauke, fl. &
fr. 12.ii.1975, *Brummitt & Lewis* 14321 (K; LISC; NDO; SRGH). S: Mazabuka, Veterinary
Research Station, fl. & fr. 6.iii.1963, *van Rensberg* 1583 (K; SRGH). **Zimbabwe**. N: Gokwe,
Copper Queen Rd., fl. & fr. 6.xi.1963, *Bingham* 897 (K; SRGH). W: Bulawayo, Norfolk Rd.,
1360 m., fl. & fr. 15.iii.1964, *Best* 397 (K; SRGH). C: Kadoma Distr., Golden Valley, fl. & fr.
14.iii.1977, *Seeds Services* VS/05 (SRGH). F: Mutare, Mutare Commonage, fl. & fr. 26.iv.1948
Chase 657 (BM; K; LISC; SRGH). **Malawi**. N: Nkhata Bay at Chikale Beach, 170 m., fl. & fr.
13.viii.1972, *Pawek* 5643 (K; SRGH). C: 6 km. N. of Nkhota Kota, fl. & fr. 16.vi.1970,
Brummitt 11442 (K). **Mozambique**. N: Mozambique Distr., Malema, Mutuali, Exp. Stat.
CICA, fl. & fr. 27.iv.1961, *Balsinhas & Marrime* 458 (COI; K; LISC; LMA). Z: Quelimane,
Mocuba, Namagoa, fl. & fr. vi.1947, *Faulkner* 160 (COI; K; SRGH). T: Tete, Boroma, Sisitso,
275 m., fl. & fr. 10.vii.1950, *Chase* 2634 (BM; K; SRGH). MS: Beira, Gorongosa National Park,
fl. & fr. viii.1970, *Tinley* 1995 (LISC; SRGH).
     Tropical Africa to S. Africa (Transvaal), Madagascar, Egypt, tropical Asia and N. Australia.
Savanna woodland, grassland, alluvial and sandy soils and cultivated ground; 0–1530 m.

2. **Ipomoea leucanthemum** (Klotzsch) Hall. f. in Engl., Bot. Jahrb. **18**: 124 (1893).—
     Dammer in Engl., Pflanzenw. Ost.-Afr. **C**: 331 (1895).—Baker & Rendle in F.T.A. **4**: 137
     (1905).—Roessler in Merxm. Prodr. Fl. SW. Afr. **116**: 15 (1967). Type: Mozambique,
     Tete, Rios de Sena, *Peters* II.1845 (B).
          Calycanthemum leucanthemum Klotzsch in Peters, Reise Mossamb., Bot. **1**: 244, t. 40
     (1861). Type as above.

A much branched prostrate, ascending or erect herb. Stems up to 80 cm. long,
slender, greyish-pubescent. Leaf lamina linear-oblong to ovate-cordate, up to 7×2·7
cm., cuneate, rounded or shallowly cordate at the base, obtuse and minutely mucronate
at the apex, thinly hairy on both surfaces; petiole short, up to 2 cm. long, pubescent.
Inflorescences axillary, 1–3 flowered, subsessile; bracts ovate-acuminate, pilose;

pedicels up to 6 mm. long, pubescent. Sepals subequal, ovate or oblong lanceolate, up to 9 mm. long, pilose. Corolla funnel-shaped, white or cream, up to 12 mm. long, midpetaline areas pilose. Ovary pilose; stigmas with two oblong lobes. Capsule globose, pilose. Seeds black, appressed pubescent.

Botswana. N. 3 km. N. of Aha Hills, fl. & fr. 13.iii.1965, *Wild & Drummond* 6964 (K; SRGH). **Zambia.** B:.Mongu, fl. & fr. 20.ii.1966, *Robinson* 6844 (K). C: Luangwa Valley Game Reserve, c. 610 m., fl. & fr. 13.ii.1967, *Prince* 179 (K; SRGH). C: Katondwe, fl. & fr. 4.ii.1964, *Fanshawe* 8300 (K; SRGH). S: Mazabuka, 6 km. from Chirundu Bridge on Lusaka Rd., fl. & fr. 6.ii.1958, *Drummond* 5491 (K; LISC; SRGH). **Zimbabwe.** N: Urungwe, N. of Mauora R., c. 610 m., fl. & fr. 26.ii.1958, *Phipps* 891a (K; LISC; SRGH). W: Hwange, Lutope, Gwaai R. Junction, fl. & fr. 26.ii.1963, *Wild* 6012 (K; LISC; SRGH). S: Bikita, Sabi Valley, Umkondo Mine, fl. & fr. 9.ii.1966, *Wild* 7515 (K; LISC; SRGH). **Malawi.** N: Karonga Distr., 8 km. W. of Karonga at Chaminade Sec. School, 550 m., fl. & fr. 15.iv.1976, *Pawek* 11051 (K; SRGH). C: Between L. Malawi and Grand Beach Hotels, nr. Salima, 480 m., fl. & fr. 16.ii.1959, *Robson & Steele* 1615 (BM; K; LISC; SRGH). S: Mangochi Distr., 26 km. NW. of Mangochi (4 km. beyond Club Makokola turn off), 500 m., fl. & fr. 20.ii.1982, *Brummitt & Patel* 15990 (K; LISC; MAL). **Mozambique.** T: N'Kanya R., opposite Msusa, fl. & fr. 25.vii.1950, *Chase* 2831 (BM; K; SRGH). GI: Inhambane, Govuro, 5 km. from Banamana to Machaíla, fl. & fr. 21.iii.1974, *Correia & Marques* 4136 (LISC).
Also in Namibia. Woodland, savanna, open bushland, grassland, roadsides and along rivers; 550–1050 m.

3. **Ipomoea ephemera** Verdc. in Kirkia **1**: 30, TAB. V. (1961). Type: Mozambique, Zambezia, *Faulkner* K423 (EA, isotype; K, holotype).

Prostrate-ascending or erect annual herb up to 1·5 m. Stems not much branched, slender, stellate-pubescent. Leaf lamina linear to linear-oblong, 1·8–6 × 0·4–1·2 cm., cuneate to shortly cordate or hastate at the base, obtuse or subacute, minutely mucronate at the apex, stellate-pubescent on both surfaces; petiole up to 7 mm. long, stellate-pubescent. Flowers axillary, solitary; bracts linear more or less 4 mm. long; pedicels 2·5 mm. long. Sepals subequal, ovate-lanceolate, up to 10 × 3 mm., membranous stellate-pubescent, ciliate at the margin. Corolla narrowly funnel-shaped, up to 3 cm. long, nearly glabrous; tube narrow, more or less 2 cm. long, purple; limb whitish. Ovary glabrous. Capsule globose, glabrous, apiculate by the style-base. Seeds black, appressed pubescent.

Mozambique. N: Nampula, between Corrane and Nampula, fl. & fr. 11.iv.1937, *Torre* 1389 (COI; LISC). Z: Lugela, Muobede Rd., fl. & fr. 19.iv.1949, *Faulkner* 423 (K, holotype; EA, isotype).
Not known elsewhere, Bushland and pans, dampish soil.

4. **Ipomoea coscinosperma** Hochst. ex Choisy in DC., Prodr. **9**: 354 (1845).—Hall. f. in Engl., Bot. Jahrb. **18**: 124 (1893).—Baker & Rendle in F.T.A. **4**: 138 (1905).—Engl. in Sitz.—Ber. Königl. Preuss. Akad. Wiss. Berl. **52**: 14 (1907).—Eyles in Trans. Roy. Soc. S. Afr. **5**: 454 (1916).—Dandy in F. W. Andr., Fl. Pl. Anglo-Egypt. Sudan **3**: 113 (1956).—Meeuse in Bothalia **6**: 721 (1958).—Verdc. in F.T.E.A., Convolvulaceae: 92 (1963).—Heine in F.W.T.A. ed. 2, **2**: 350 (1963).—Roessler in Merxm. Prodr. Fl. SW. Afr. **116**: 14 (1967).—Ross, Fl. Natal: 295 (1972). Type from Sudan Republic.
Ipomoea coscinosperma Choisy var. *glabra* Rendle in F.T.A. **4**: 138 (1905). Type from Ethiopia.
Ipomoea coscinosperma Choisy var. *hirsuta* A. Rich., Tent. Fl. Abyss. **2**: 66 (1851). Type from Ethiopia.
Ipomoea polygonoides Schweinf., Beitr. Fl. Aethiop.: 95 (1867). Type from Ethiopia.

Spreading annual, glabrescent or hairy. Stems several, more or less stout, often angular, thinly hairy, glabrescent, at first suberect but soon prostrate. Leaf lamina from narrowly linear lanceolate to oblong, 3–10 × 1–4 cm., entire or somewhat repand, blunt but apiculate at the apex, rounded cuneate or shallowly cordate at the base, glabrescent, pilose or hairy; petiole 5–25 mm. long, more or less pilose. Inflorescence axillary, subsessile, usually 1–3 flowered, pilose; peduncles and pedicels up to 5 mm. long; bracts linear-subulate, about 4 mm. long, pilose. Sepals subequal, lanceolate or ovate up to 8 mm. long, acuminate at the apex with hyaline lower margins, pilose. Corolla narrowly funnel-shaped, red or white, slightly longer than the calyx. Capsule globose, glabrous, crowned by the persistent style base. Seeds brown very shortly pubescent.

Botswana. N: Tsantsarra Pan, Chobe National Park, fl. & fr. 22.i.1978, *Smith* 2196 (SRGH). SE: 60 km. NW. of Serowe, fl. & fr. 24.iii.1965, *Wild & Drummond* 7290 (K; LISC; SRGH). C: Mfuwe, fl. & fr. 16.v.1969, *Astle* 5697 (K; LISC; SRGH). S: Kafue Flats, fl. & fr. 17.vi.1956, *Angus* 1537 (K; LISC; SRGH). **Zimbabwe**. W: Bulawayo, Norfolk Rd., 1360 m., fl. & fr. 22.iii.1964, *Best* 401 (K; SRGH). C: Chegutu, cult. at Harare Agric. Res. Sta., fl. & fr. 7.ii.1978, *Seed Services* VS/74 (SRGH). E: Chipinge, Lower Sabi, East Bank, nr. Hippo Mine, 365 m., fl. & fr. 12.iii.1957, *Phipps* 587 (SRGH). S: Gwanda 608 m., fl. & fr. v. 1955, *Davies* 1354 (K). **Mozambique**. GI: Gaza, from Chibuto to Caniçado, between Sousuaningue and Caniçado, fl. & fr. 2.vi.1959, *Barbosa & Lemos* 8570B (COI; K; LISC; SRGH). M: Magude, nr. Mapulanguene, fl. & fr. 23.i.1948, *Torre* 7192 (C; LISC; LMA; MO; UPS; WAG).

Eastern Africa from Sudan Republic, Ethiopia to S. Africa, W. Africa and Namibia. Damp sand of riverine areas and cultivated ground; 365–1360 m.

5. **Ipomoea plebeia** R. Br., Prodr. Fl. Nov. Holl., ed. 1: 484 (1810).—van Ooststr. in Fl. Males., Ser. 1, **4**: 463 (1953).—Meeuse in Bothalia **6**: 723 (1957).—Verdc. in Kew Bull. **13** (2): 199 (1958); in F.T.E.A., Convolvulaceae: 92, fig. 21 (1963).—Roessler in Merxm. Prodr. Fl. SW. Afr. **116**: 16 (1967).—Ross, Fl. Natal: 296 (1972).—Compton, Fl. Swaziland: 478 (1976). Type from Australia.

Annual with twining or prostrate, slender, patently pubescent stems. Leaf lamina triangular-ovate, 4–9 × 3–5·5 cm., entire, cordate at base with the basal lobes rounded, acuminate at the mucronate apex, sparsely pilose and ciliate at the margin; petiole 1·5–5 cm. long, slender, pilose to hispidulous like the stems. Flowers axillary, solitary or in 2–5 flowered cymes, sessile or with peduncle up to 10 mm long; pedicels 3–8 mm. long; bracteoles minute, lanceolate, up to 2·5 mm. long, pilose. Sepals unequal, herbaceous to subcoriaceous, ovate triangular, acute or acuminate at the apex, contracted or rounded below, hairy to glabrescent, accrescent in fruit; the outer three 4–6·5 mm. long, with ciliate margins; the inner two linear-oblong, more or less 4 mm. long, less ciliate. Corolla funnel-shaped, white or with purple centre, 6–9 mm. long, midpetaline areas distinct, pilose outside mainly towards the tips. Ovary glabrous, style short; stigmas purplish. Capsule ovoid, glabrous, cuspidate by the persistent stylebase. Seeds subtrigonous, shortly brownish or greyish-pubescent.

Subsp. **africana** Meeuse in Bothalia **6**: 723 (1958).—Verdc. in F.T.E.A., Convolvulaceae: 94, fig. 21 (1963).—Roessler in Merxm. Prodr. Fl. SW. Afr. **116**: 16 (1967).—Ross, Fl. Natal: 296 (1972). Type from S. Africa (Transvaal).

Ipomoea cardiosepala Baker & Wright in Dyer, F.C. **4**: 62 (1904) non Meisn. Type from S. Africa.

Ipomoea cynanchifolia sensu Baker & Rendle in F.T.A. 4: 137 (1905) non (Roxb.) C.B. cl., nec Meisn. Type from the Himalaya.

Sepals sparsely pilose; corolla white with purple centre.

Botswana. N: Maun, residential area, 930 m., fl. & fr. 11.iii.1972, *Biegel & Russel* 3926 (K; LISC; SRGH). SE: Mochudi, Phutodikobo Hill, 914–1066 m., fl. & fr. 14.iii.1967, *Mitchison* 47 (K). **Zambia**. B: Masese, fl. & fr. 14.iii.1961, *Fanshawe* 6432 (SRGH). W: Kitwe, fl. & fr. 16.ii.1971, *Fanshawe* 11144 (K; SRGH). C: Lusaka, fl. & fr. 15.iii.1957, *Noak* 149 (K; SRGH). E: Chipata Distr., c. 32 km: SE of Mfuwe, 600 m., fl. & fr. 28.ii.1969, *Astle* 5554 (K; SRGH). S: between Mazabuka and Magoye, fl. & fr. 28.ii.1963, *van Rensberg* 1535 (K; SRGH). **Zimbabwe**. N: Binga, Mwenda Research Station, fl. & fr. 3.vi.1966, *Grosvenor* 81 (K; LISC; SRGH). W: Hwange, Makwa Pan, 14 km. SE. of Main Camp, Hwange National Park, fl. & fr. 14.iv.1972, *Grosvenor* 694 (K; LISC; SRGH). C: Chegutu, Poole Farm, 1220 m., fl. & fr. 3.iv.1946, *Wild* 1012 (K; SRGH). E: Mutare Commonage, 1100 m., fl. & fr. 6.iii.1956, *Chase* 6006 (BM; K; SRGH). S: Mwenezi, Malangwe R., SW. Meteke Hills, 760 m., fl. & fr. 6.v.1958, *Drummond* 5607 (K; LISC; SRGH). **Malawi**. N: Karonga Distr., Vinthukhutu Forest, c. 3 km. N. of Chilumba, fl. & fr. 13.iv.1976, *Pawek* 10972 (K). C: Lilongwe, 1050 m., fl. & fr. 23.ii.1970, *Brummitt & Little* 9328 (K; LISC; MA1; PRE; SRGH). S: Lake Malawi, Boadzulu Isl., fl. & fr. 14.iii.1955, *Exell, Mendonça & Wild* 881 (BM; SRGH). **Mozambique**. T: Tete, between Estima and Marueira, c. 10 km. from Estima, fl. & fr. 10.iii.1972, *Aguiar Macedo* 5038 (LISC; LMA; LMU). GI: Gaza, from Chibuto to Caniçado, fl. & fr. 2.vi.1959, *Barbosa & Lemos* 8570A (LMA). M: Maputo, Goba, Umbeluzi R., fr. 19.iv.1949, *Myre & Balsinhas* 670 (LMA).

Also in Angola, Namibia, S. Africa (Transvaal and Natal). Woodland, dry bush, grassland, alluvial zones, roadside and cultivated land; 600–1525 m.

Subsp. *plebeia* occurs in Australia and Malaysia. It has laxly pubescent sepals and is distinctly more hairy than subspec. *africana*.

6 **Ipomoea gracilisepala** Rendle in Journ. Bot., Lond. **39**: 12 (1901).—Baker & Wright in
Dyer, F.C. **4**: 58 (1904).—Meeuse in Fl. Pl. Afr. 31, pl. 1217b (1956); in Bothalia **6**: 725
(1958). Type from S. Africa.
 Ipomoea xiphosepala Baker in Dyer, F.C. **4**: 58 (1904) non Baker in Kew Bull. **1894**: 69
(1894). Type from S. Africa (Transvaal).
 Ipomoea lyciifolia Merxm. in Trans. Rhod. Sci. **43**: 40 (1951). Type: Zimbabwe,
Marondera, *Dehn* 786 (K, holotype; SRGH, isotype).

Procumbent or trailing herb. Stems several from the base, up to 1·10 m. long,
subterete, at first densely, later finely and shortly pubescent. Leaf lamina oblong or
lanceolate-oblong to lanceolate, 2–7 × 0·5–1 cm., usually hastate-truncate or auricled
at the base, sometimes some lanceolate from a narrow base or all lanceolate, obtuse,
minutely mucronate at the apex, entire to subrepand, hairy or glabrescent; petiole 5–10
mm. long, glabrescent; inflorescences axillary 1–2 flowered; peduncle absent or very
short, thinly hairy with rather long hairs as are bracteoles, pedicels and sepals;
bracteoles linear-lanceolate, 3–6 mm. long, obtuse or subacute; pedicels up to 10 mm.
in fruit. Sepals subequal, lanceolate to linear, acuminate, 7–15 × 1–2 mm., spreading
in fruit. Corolla narrowly funnel-shaped, mauve or white, 12–15 mm. long; midpetaline areas hairy towards the tips. Capsule globose, hirsute. Seeds brown with a villous
tomentum of appressed stiff hairs.

 Zimbabwe. W: Matobo, Farm Besna Kobila, 1435 m., fl. & fr. iii.1954, *Miller* 2248 (K;
SRGH). C: Harare, fl. & fr. iii.1919, *Eyles* 1546 (BM; PRE; SRGH).
 Also in S. Africa (Transvaal and Orange Free State); 1435–1530 m. Grassland and damp land
near rivers.

7. **Ipomoea polymorpha** Roem. & Schult., Syst. **4**: 254 (1819).—van Ooststr. in Fl. Males.,
Ser. 1, **4**: 464, fig. 38, 39 (1953).—Verdc. in Kew Bull. **13**: 200 (1958); in F.T.E.A.,
Convolvulaceae: 94 (1963). Type from Australia.
 Ipomoea heterophylla R. Br., Prodr.: 487 (1810) nom. illegit. non Ortega (1800).—Peter
in Engl., Bot. Jahrb. **18**: 125 (1893). Type as above.
 Ipomoea commatophylla Steud. ex A. Rich., Tent. Fl. Abyss. **2**: 65 (1851).—Baker &
Rendle in F.T.A. **4**: 139 (1905). Type from Ethiopia.

Annual or biennial. Stems erect or prostrate, simple or branched from the base;
young parts densely pilose, adult parts less densely so to glabrous. Leaf lamina ovate,
oblong or elliptic, 2–4 × 1–1·6 cm., base acute, attenuate into the petiole, apex acute or
obtuse to rounded, mucronulate, margin subentire, undulate or coarsely dentate,
shallowly to deeply lobed, sometimes irregularly pinnatifid with few segments, or lyrate
with a large, ovate or elliptic entire to coarsely dentate terminal segment and small
triangular to hastate basal ones, pubescent; petiole up to 2 cm. long, sparsely pilose.
Flowers axillary, solitary; peduncle and pedicel very short or absent; bracts linear, up
to 7 mm. long, hairy. Sepals unequal, the outer three more or less 11 mm. long, pilose,
with the ovate-circular basal part subhyaline and coriaceous, with prominent foliaceous
midrib and margins and a narrow attenuate acute apical part, the inner sepals narrower.
Corolla narrowly funnel-shaped, white or pink often with a darker centre, 1·4–1·6 cm.
long. Ovary glabrous. Capsule globose, glabrous, straw-coloured. Seeds densely
appressed pubescent.

 Zambia. N: Mbala Distr., Kawimbi to M'panda, 1525 m., fl. & fr. 11.ii.1955, *Richards* 4456
(K).
 Also in Ethiopia, Uganda, Kenya, Tanzania, India, China, Malaysia and Australia. Grassland;
1525 m.

8. **Ipomoea kotschyana** Hochst. ex Choisy in DC., Prodr. **9**: 354 (1845).—Hall. f. in Engl.,
Bot. Jahrb. **18**: 125 (1893).—Baker & Rendle in F.T.A. **4**: 139 (1905).—Dandy in F.W.
Andr., Fl. Pl. Anglo-Egypt. Sudan **3**: 114 (1956).—Verdc. in Kew Bull. **13**: 200 (1958);
Kew Bull. **15**: 6 (1961); F.T.E.A., Convolvulaceae: 95 (1963).—Heine in F.W.T.A. ed. 2,
2: 351 (1963). Type from Sudan Republic.
 Ipomoea laciniata Balf. f. in Proc. Roy. Soc. Edinb. **12**: 82 (1882); Trans. Roy. Soc.
Edinb. **31**: 188 (1888). Type from Socotra.

Annual or biennial herb with long slender rhizome. Stems very slender, trailing,
pubescent to glabrescent. Leaf lamina lanceolate to oblong-ovate in outline, 1·5–3 ×
0·4–2 cm.; upper leaves lanceolate, entire or with two small basal lobes; lower and

radical leaves 1–2-pinnatifid with the lobes linear 0·5–1·5 × 0·1–0·3 cm., sometimes sparsely dentate, pubescent; petiole 0·2–1·5 cm. long. Flowers axillary, solitary, subsessile; pedicels up to 3 mm. long. Sepals unequal, pubescent, the outer three 6 mm. long, base ovate and coriaceous, often dentate on either side, apex narrow; the two inner ones much shorter, elliptic, hyaline. Corolla very narrowly tubular, pink with dark centre, more or less long, opening in the evening. Capsule globose, glabrous, 6-valved. Seeds 4–6, small collateral, together forming a sphere, pale brown with dark spots and patches or pubescence.

Mozambique. MS: Beira, fl. & fr. 5.xi.1898, *Braga* s.n. (COI).

Also in Mali, Sudan Republic, Ethiopia, Kenya, Tanzania and Socotra. Grassy banks on sandy soil.

9. **Ipomoea oenotherae** (Vatke) Hall. f. in Engl., Bot. Jahrb. **18**: 125 (1893).—Dammer in Engl., Pflanzenw. Ost.-Afr. **C**: 331 (1895).—Baker & Wright in Dyer, F.C. **4**, 2: 49 (1904).—Baker & Rendle in F.T.A. **4**: 145 (1905).—Meeuse in Bothalia **6**: 727 (1958). —Verdc. in Kew Bull. **13**: 200 (1958); F.T.E.A., Convolvulaceae: 95 (1963).—Roessler in Merxm. Prodr. Fl. SW. Afr. **116**: 16 (1967).—Binns, H.C.L.M.: 40 (1968).—Ross, Fl. Natal: 296 (1972).—Jacobsen in Kirkia **9**: 171 (1973). Type from Kenya.

Convolvulus oenotherae Vatke in Linnaea **43**: 520 (1882). Type as above.

Perennial, with a fusiform tuberous rootstock. Stems numerous, slender, up to 50 cm. long, prostrate or ascending, densely white-pilose above, glabrescent beneath. Leaves very variable, the basal ones rosulate, (if present) long petiolate with a petiole up to 5 cm. long); leaf lamina lanceolate or linear, up to 8 × 1 cm., entire or with 1–2 lateral teeth or lobes or repando-pinnatisect 2–7-lobed, pilose, the apical lobe usually longer; cauline leaves 2–6·5 cm. long, usually subpalmately or pinnately 3–7-lobed, with linear to lanceolate lobes, the middle ones usually longest and often lobed or dentate at first with soft silvery-white hairs, soon glabrescent, main nerves prominent beneath. Flowers axillary, solitary, subsessile or with a pedicel up to 15 mm. long; bracteoles linear-subulate or linear-filiform, 10–15 × 0·5 mm., pubescent. Sepals unequal ovate or ovate-lanceolate, aristate, ciliate with median keel of hairs, outer ones herbaceous, up to 13 mm. long, inner ones paler, more membranous and shorter, all becoming broader and brown in fruit. Corolla narrowly funnel-shaped with spreading limb, mauve to purple, 2·3–5 cm. long, midpetaline areas glabrous. Capsule globose, glabrous, straw-coloured. Seeds densely appressed pubescent.

Var. **oenotherae**

Ipomoea petuniodes Baker in Dyer, F.C. **4**: 63 (1904). Type from S. Africa (Transvaal).
Ipomoea cecilae N.E. Br. in Kew Bull. **1906**: 166 (1906). Type: Zimbabwe, Mutare, *Cecil* 36 (K, holotype).
Ipomoea pachypus Pilger in Engl., Bot. Jahrb. **41**: 296 (1908). Type from Ethiopia.
Ipomoea lineariloba Chiov. in Miss. Biol. Borana, Angiosp.—Gymnosp.: 174, fig. 51 (1939). Type from Ethiopia.

Corolla up to 5 cm. long. Leaves not as narrow as in var. *angustifolia*.

Botswana. N: Ramokgwebana area nr. Tsessebe Railway Station, fl. & fr. 5.i.1974, *Ngoni* 239 (K; SRGH). SE: c. 56 km. N. of Pilane, Bodumgwane, 1040 m., fl. & fr. 18.iii.1967, *Mitchinson* 99 (K). **Zambia.** W: Luanshya, fl. & fr. 18.i.1955, *Fanshawe* 1802 (K). S: Mazabuka, Vet. Research Station, fl. & fr. 29.xii.1962, *van Rensberg* 1152 (K; SRGH). **Zimbabwe.** N: Mazoe, Umvukwes, Ruorka Ranche, 1525 m., fl. & fr. 16.xii.1952, *Wild* 3923 (K; SRGH). W: Matobo, fl. & fr. 11.xii.1947, *West* 2485 (K; SRGH). C: Hunyani, Old Charter Rd., 1220 m., fl. & fr. 27.xii.1926, *Eyles* 4590 (K; SRGH). **Malawi.** S: Ntcheu Distr., Chirobwe 1580 m., fl. & fr. 18.iii.1955, *Exell, Mendonça & Wild* 1020 (BM; SRGH).

Also Ethiopia, Somali Republic, Uganda, Kenya, Tanzania, Namibia and S. Africa (Transvaal). Mixed woodland, grassland and sandy soil; 1000–1580 m.

The var. *angustifolia* (Oliv.) Verdc. has corolla mostly 1·4–1·6 cm. long and leaves narrow with narrow lobes 2–4 mm. wide. It occurs in Uganda and Kenya.

10. **Ipomoea hackeliana** (Schinz) Hall.f. in Engl., Bot. Jarhb. **18**: 126 (1893).—Baker & Rendle in F.T.A. **4**, 2: 146 (1905).—Meeuse in Bothalia **6**: 726 (1958).—Roessler. in Merxm. Prodr. Fl. SW. Afr. **116**: 14 (1967). Type from Namibia.

Aniseia hackeliana Schinz in Verhandl. Bot. Ver. Brandenb. **30**: 274 (1888). Type as above.

Annual. Stems several from the base, up to 2 m. long, prostrate, with soft patent hairs when young. Leaf lamina ovate-cordate, 1–4 × 0·75–3 cm., more or less distinctly crenate or crenate-dentate, acute to rounded at the apex, sparsely pilose on both surfaces, pellucidly glandular when dry; basal sinus usually wide and shallow with the blade cuneately decurrent into the slender up to 2 cm. long pilose petiole. Flowers solitary or in few-flowered fascicles; pedicels slender, pilose, up to 15 mm. long; bracteoles subcordate at the base, acute, ciliate or midrib and margin, up to 3 mm. long. Sepals unequal, up to 10 mm. long, green often suffused with purplish red, pilose, the outer three ovate from a cordate base, obtuse or subacute, the two inner lanceolate, acute, all accrescent, becoming papyraceous with distinct finely raised veins, attaining 13 × 9 mm. Corolla funnel-shaped, white or mauve, 10–13 mm. long; midpetaline areas slightly hairy. Ovary hairy. Capsule subglobose, densely pilose. Seeds black, hairy.

Botswana. SW: Kgalagadi, c. 100 km. NW. of Tshabong, 975 m., fl. & fr. 1.iii.1963, *Leistner* 3118 (K; PRE). SE: Khutse, 915 m., fl. & fr. 20.iv.1972, *Coleman* 412 (K; SRGH). **Zimbabwe**. S: Beitbridge, Chiturupazi, fl. & fr. 18.iii.1967, *Drummond* 9026 (SRGH).
Also Namibia and S. Africa (Transvaal). Open woodland, savanna, sandy soil and roadside; 900–1000 m.

11. **Ipomoea recta** De Wild. in Ann. Mus. Congo Belge, Bot. Ser. **4**: 114 (1903).—Baker & Rendle in F.T.A. **4**, 2: 141 (1905).—Verdc. in Hook., Ic. Pl. **36**: t. 3556 (1956); in Kew Bull. **33**: 164 (1978). Type from Zaire.
Ipomoea debeerstii De Wild. in Ann. Mus. Congo Belge, Bot. Ser. **4**: 114 (1903).—Baker & Rendle in Dyer, F.T.A. **4**, 2: 142 (1905). Type from Zaire.

Perennial herb. Stems several, from a woody rootstock, radiating, erect or prostrate, unbranched, up to 45 cm. long, striate, densely covered with short spreading hairs. Leaf-lamina linear-lanceolate or oblong, 20–50 × 2–10 mm., base cuneate or rounded, apex apiculate, upper surface thinly hairy, lower surface with a silvery indumentum, margin ciliate; petiole 0–3 mm. long. Flowers solitary, axillary, erect; peduncle up to 10 mm. long bearing above the middle 2-linear-lanceolate persistent 6–12 mm. long bracteoles. Sepals ovate-lanceolate, apex acute or acuminate, up to 8 × 3 mm. long, hairy outside, margin ciliate, glabrous inside. Corolla tubular funnel-shaped, white or pink with purple centre, up to 4·5 cm. long; tube slightly swollen at the middle. Stamens hairy at the base. Ovary glabrous. Capsule globose, glabrous. Seeds more or less globose, velutinous tomentose.

Zambia. N: Kawambwa Distr., M'tunatusha R., 1290 m., fl. & fr. 28.xi.1961, *Richards* 15423 (K; LISC). W: Mwinilunga Distr., Kalenda Plain N. of Matonchi Farm, fl. & fr. 8.xii.1937, *Milne-Redhead* 3560 (BM; K). E: Sesare, Copper Mine, 700 m., fl. & fr. 9.xii.1958, *Robson* 876 (BM; K; LISC; SRGH). **Malawi**. N: Mzimba Distr., c. 32 km. W. of Mzuzu, Lunyangwa R., bridge, 1160 m., fl. & fr. 2.iii.1974, *Pawek* 8155 (K).
Also Zaire and Tanzania. Woodland, savanna, grassland and sandy soil; 700–1300 m.

12. **Ipomoea milnei** Verdc. in Hook. Ic. Pl. **36** t. 3557 (1956). Type from Angola.

Perennial herb. Stems generally few branched, prostrate, very slender 10–120 cm. long, obscurely striate, radiating, more or less pubescent. Leaf-lamina narrowly linear, 10–70 × 0·5–0·8 mm., more or less sessile and erect, obtuse at the apex, cuneate at the base, glabrous or more or less pubescent mainly on the central nerve. Flowers axillary, solitary, buds with apical hair tufts; peduncle 2–25 mm. long, pedicels 4–11 mm. long; bracts linear up to 8 mm. long. Sepals lanceolate or ovate-lanceolate, narrowly and long acuminate, 8–10 mm. long, glabrous or more or less pubescent, the inner ones with hyaline edges. Corolla funnel-shaped, with cylindric tube, pale mauve, deeper inside throat, rarely reported to be primrose yellow, up to 2·6 cm. long. Capsule globose, glabrous. Seeds greyish, very shortly puberulous.

Zambia. N: Kasama, Misambu, 1341 m., fl. & fr. 20.i.1962, *Astle* 1275 (K; SRGH).
Also in Angola. Sandy and rocky soils; 1320–1341 m.

13. **Ipomoea blepharophylla** Hall. f. in Engl., Bot. Jahrb. **18**: 125 (1893).—Baker & Rendle in F.T.A. **4**, 2: 141 (1905).—Engl. in Sitz.-Ber. Konigl. Preuss. Akad. Wiss. Berl. **52**: 24 (1907).—Fries, Wiss. Ergebn. Schwed. Rhod.-Kongo-Exped.: 269 (1916).—Eyles in

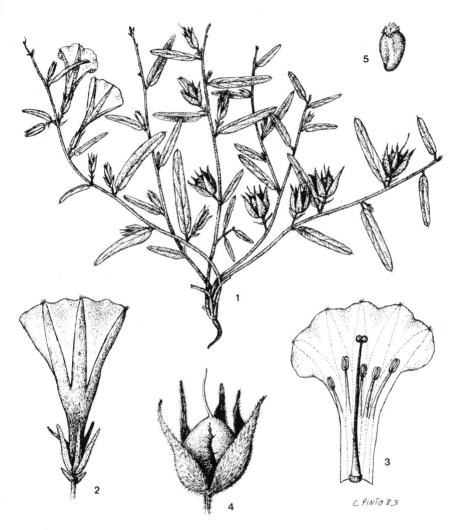

Tab. 17. IPOMOEA BLEPHAROPHYLLA. 1, habit ($\times\frac{1}{2}$); 2, flower ($\times$1); 3, corolla opened to show stamens and pistil ($\times$1); 4, fruit ($\times$2); 5, seed ($\times$2). All from *Drummond* 4906.

Trans. Roy. Soc. S. Afr. **5**: 453 (1916).—Wild, Guide Flora Victoria Falls: 155 (1953). —Brenan in Mem. N.Y. Bot. Gard. **9**: 7 (1954).—Dandy in F. W. Andr., Fl. Pl. Anglo-Egypt. Sudan **3**: 120 (1956).—Verdc. in F.T.E.A., Convolvulaceae: 96 (1963). —Heine in F.W.T.A. ed. 2, **2**: 350 (1963).—Binns, H.C.L.M.: 40 (1968).—Jacobsen in Kirkia **9**: 171 (1973). TAB. **17**. Type from Sudan.

Ipomoea blepharophylla var. *cordata* Rendle in Journ. of Bot. **34**: 37 (1896).—Baker & Rendle in F.T.A. **4**, 2: 141 (1905). Type from Kenya.

Ipomoea glossophylla Chiov. in Ann. Bot. Roma **9**: 83 (1911). Type from Ethiopia.

Perennial herb. Stems several, from a woody rootstock, prostrate, densely clothed with short yellowish spreading hairs. Leaf lamina lanceolate or narrowly oblong, up to 8 × 2·5 cm., obtuse or mucronate at the apex, rounded or subcordate to minutely cordate at the base, glabrous or with some hairs on the midrib above, the margins and veins beneath ciliate; petiole up to 7 mm. long, pubescent. Flowers solitary, axillary; peduncle up to 12 mm. long; pedicels up to 20 mm. long; bracteoles lanceolate, 3·5 mm. long. Sepals lanceolate, up to 18 × 4 mm., pubescent. Corolla tubular-funnel-shaped, distinctly narrowed below, mauve with darker centre or whitish, up to 6 cm. long, the tube sometimes 2 cm. long. Capsule globose, glabrous, tipped by persistent style-base. Seeds brownish with appressed yellowish hairs.

Zambia. B: Masese, fl. & fr. 13.ix.1969, *Mutimushi* 3646 (SRGH). N: Mbala Distr., Chilongwelo, 1495 m., fl. 4.i.1952, *Richards* 385 (K). W: Lake Ishika, fl. & fr. 9.viii.1953, *Fanshawe* 212 (K). C: Kabwe, fl. & fr. 30.x.1963, Fanshawe 8009 (K). S: Mapanza Mission, 1070 m., fl. & fr. 29.xii.1952, *Robinson* 31 (K). **Zimbabwe**. N: Lomagundi, 1174 m. fl. & fr. 9.x.1962, *Jacobsen* 1806 (PRE). W: Bulawayo, 1220 m., fl. & fr. 14.ii.1965, *Best* 417 (K; SRGH). C: Harare, between Avonvale West and Mabelreign, 1480 m., fl. & fr. 16.x.1955, *Drummond* 4906 (K; LISC; SRGH). E: Inyanga Distr., Honde Valley, 762 m., fl. & fr. 7.iii.1966, *Plowes* 2762 (K; LISC; SRGH). S: Masvingo, Mushandike National Park, fl. & fr. 26.ii.1974, Bezuidenhout 135 (K; SRGH). **Malawi**. C: Nkhota Kota, 1600 m., fl. & fr. 12.ii.1944, *Benson* 750 (PRE). **Mozambique**. T: Macanga, Angónia Rd., 15 km. from Furancungo, fl. & fr. 29.ix.1942, *Mendonça* 521 (LISC; LMA; WAG).

Also in Sudan, Ethiopia, Uganda, Kenya, Tanzania, West Africa and Angola. Woodland, grassland, sandy soil; 760–1500 m.

14. **Ipomoea fulvicaulis** (Hochst. ex Choisy) Boiss. ex Hall. f. in Engl., Bot. Jahrb. **18**: 128 (1893).—Britten, Baker, Rendle & al. in Trans. Linn. Soc., London 2 Ser., **4**: 29 (1894).—Baker & Rendle in F.T.A. **4**, 2: 143 (1905).—Verdc. in Kew Bull. **14**: 338 (1960); in F.T.E.A., Convolvulaceae: 97, fig. 22, 6 (1963).—Binns, H.C.L.M.: 40 (1968).—Jacobsen in Kirkia **9**: 171 (1973). Type from Ethiopia.

Perennial herb with woody rootstock. Stems up to 2 m. long, twining or erect, slender, persistently yellow-brown-pubescent. Leaf lamina narrowly oblong, ovate or elliptic-ovate, 3–11·5 × 1–4·5 cm., obtuse or emarginate at the apex, cordate to truncate at the base, pubescent above and densely so beneath; petiole very hairy, short, 1–4·5 cm. long. Flowers axillary, solitary or a few germinate in small heads, peduncles 1–7·5 cm. long, hairy; bracts linear to ovate, acuminate, 5–19 × 1–8 mm. Sepals lanceolate to ovate, acute, up to 16 × 6 mm. Corolla funnel-shaped, mauve or purple, more or less pilose, 2·5–5 cm. long. Capsule globose, glabrous. Seeds brownish, pubescent.

1. Bracts ovate, acuminate, up to 19 × 8 mm. long; sepals ovate; leaves generally more or less densely pubescent   -   -   -   -   -   -   -   -   -   var. *fulvicaulis*
    — Bracts linear to lanceolate, up to 23 × 6 mm.; sepals ovate or lanceolate; leaves covered with stiff yellowish hairs, pubescent or glabrescent   -   -   -   -   -   -   2
2. Bracts linear, up to 7 mm. long; sepals ovate or lanceolate; leaves narrowly oblong to ovate-oblong, covered on both surfaces with short rather stiff yellowish hairs
    var. *asperifolia*
    — Bracts lanceolate, up to 15 × 4 mm.; sepals evidently unequal, the outer ones with an ovate basal part up to 9 × 6 mm., and a lanceolate apical part up to 12 × 2·5 mm., the inner ones narrower; leaves narrowly oblong, pubescent and glabrescent -   -   -   var. *heterocalyx*

Var. **fulvicaulis** Verdc. in Kew Bull. **14**: 338 (1960); in F.T.E.A., Convolvulaceae: 97, fig. 22/6 (1963). Type from Ethiopia.
    Aniseia fulvicaulis Hochst. ex Choisy in DC., Prodr. **9**: 431 (1845).—A Rich., Tent. Fl. Abyss. **2**: 74 (1851). Type as above.

Botswana. N: Chobe National Park, between Serondella and Ngwezumba R., fl. 17.x.1972, *Pope, Biegel & Russell* 808 (K; SRGH). **Zambia**. N: Mbala Distr., Mbala-Kambole Rd., 1500 m., fl. & fr. 10.ix.1960, *Richards* 13212 (K). W: Solwezi, fl. & fr. 21.ix.1930, *Milne-Redhead* 1163 (K). **Zimbabwe**. W: Matobo, Farm Besna Kobila, 1·465 m., fl. & fr. iii.1960, *Miller* 7195 (SRGH). C: KweKwe-Harare roadside, 1372 m., fl. & fr. ii.1960, *Davies* 2756 (SRGH). E: Mutare, 1220 m., fl. & fr. 6.ii.1934, *Davies* (BM). S: Gwanda, Iuli Experimental Station, fl. & fr. 9.xii.1964, *Norris-Rogers* 446 (SRGH). **Malawi**. S: Mulanje Mt., 1891, *Whyte* (BM). **Mozambique**. N: Maniamba, between Vila Cabral and Maniamba, 1400 m., fl. & fr. 29.ii.1964, *Torre & Paiva* 10894 (LISC).

Also in Ethiopia, Zaire, Kenya and Tanzania. Woodland, open forest and roadside; 1200–1500 m.

Var. **asperifolia** (Hall. f.) Verdc. in Kew Bull. **14**: 338 (1960); in F.T.E.A., Convolvulaceae: 97 (1963). Type from Angola.
    Ipomoea asperifolia Hall. f. in Engl., Bot. Jahrb. **18**: 128 (1893).—Baker & Rendle in F.T.A. 4, **2**: 143 (1905). Type as above.

Zambia. W: Solwezi Dambo, fl. & fr. 10.ix.1952, *White* 3208 (FHO; K). S: Mukwela, fl. & fr.

18.vi.1920, *Rogers* 26049 (K). **Zimbabwe**. N: Umvukwes, Darwin, Musengezi R., 1370 m., fl. & fr. 23.xii.1952, *Wild* 3987 (K; LISC; SRGH). C: Harare, fl. & fr. 19.i.1929, *Eyles* 6996 (K; SRGH). E: Chimanimani Mt., Martin Forest Reserve, fl. & fr. 14.xi.1967, *Mavi* 609 (K; SRGH). **Malawi**. N: Karonga Distr., c. 3·2 km. N. of Chilumba, Vinthukhutu Forest Reserve, 550 m., fl. & fr. 1.i.1973, *Pawek* 6288 (K; SRGH). C: Nkhota Kota, fl. & fr. 29.ii.1944, Benson 756 (PRE). S: Mulanje Mt., foot of Litchenya Path, 1040, m., fl. & fr. ii.1982 *Brummitt & Polhill* 15964 (K; MAL; LISC). **Mozambique**. MS: Manica, Moribane, *Salbany* 13 (LISC).

Also in Tanzania and Angola. Woodland, savanna bush, grassland, granite sand and roadside; 550–1500 m.

Var. **heterocalyx** (Schulze-Menz) Verdc. in Kew Bull. **15**: 7 (1961); in F.T.E.A., Convolvulaceae: 97 (1963).—Jacobsen in Kirkia **9**: 171 (1973). Type from Tanzania.

    Ipomoea heterocalyx Schulze-Menz in Notizbl. Bot. Gart. Berl. **14**: 110 (1938).

    Ipomoea dehniae Merxm. in Trans. Rhod. Sci. Assoc. **43**: 39 (1951). Type: Zimbabwe, Marondera, *Dehn* 196 (M, holotype, BR, K, isotypes).

    Zambia. N: Mbala Distr., Kaniyka Flats, path to Chinakila, 1200 m., fl. & fr. 15.i.1965, *Richards* 19564 (K). W: Mwinilunga Distr., nr. Kalenda, fl. & fr. 8.x.1938, *Milne-Redhead* 2637 (BM; K; LISC). E: Lukusuzi, 1036 m., fl. & fr. 16.viii.1970, *Sayer* 699 (SRGH). S: Mumbwa, fl. & fr. vi.1912, *Macaulay* (K). **Zimbabwe**. N: Lomagundi, Sinoia-Sanyati Rd., 1067 m., fl. & fr. 19.xii.1965, *Jacobsen* 2804 (PRE; SRGH). W: Hwange, Kazuma Forest, 918 m., fl. & fr. 25.i.1974, *Gonde* 17/74 (BM; SRGH). C: Harare, 1465 m., fl. & fr. 12.vii.1931, *Brain* 5469 (K). E: Inyanga Distr., Inyanga stream, 760 m., fl. & fr. 7.iii.1966, *Chase* 8416 (SRGH). **Malawi**. N: Nkhata Bay Distr., SW. of Chikangawa 1740 m., fl. & fr. 17.viii.1978, *Phillips* 3772B (K). C: Dowa Distr., Kongwe Forest Reserve, 1500 m., fl. & fr. 7.iii.1982, *Brummitt, Polhill & Banda* 16375 (K). S: Blantyre Distr., Ndirande Forest Reserve, 3 km. N. of Limbe, 1250 m., fl. & fr. 15.ii.1970, *Brummitt* 8580 (K). **Mozambique**. M: Maputo, between Matola and Umbeluzi, fl. & fr. 2.xii.1947, *Barbosa* 617 (C; LISC; LMA; WAG).

Also in Tanzania and Angola. Woodland, open forest, grassland, bushland and roadside weed; 350–1750 m.

15. **Ipomoea crassipes** Hook. in Bot. Mag. **70**: t. 4068 (1844).—Hiern, Cat. Afr. Welw. **1**: 732 (1898).—Hall. f. in Bull. Herb. Boiss. **7**: 44 (1899).—Rendle in Journ. Bot., **39**: 14 (1901).—Baker & Wright in Dyer, Fl. Cap. **4**: 56 (1904).—Baker & Rendle in F.T.A. **4**, 2: 147 (1905).—Suesseng. & Merxm. in Trans. Rhod. Sc. Assoc. **43**: 38 (1951).—Meeuse in Bothalia **6**: 730 (1958).—Verdc. in F.T.E.A., Convolvulaceae: 98 (1963).—Roessler in Merx. Prodr. Fl. SW. Afr. **116**: 14 (1967).—Binns, H.C.L.M.: 40 (1968).—Ross, Fl. Natal: 295 (1972).—Jacobsen in Kirkia **9**: 171 (1973).—Compton, Fl. Swaziland: 477 (1976). Type from S. Africa.

A very polymorphic perennial herb. Stems several, from a woody fusiform rootstock, erect or creeping, up to 1·20 m. long, slender, densely and softly hirsute or occasionally glabrous, as are all other vegetative parts, peduncles, pedicels, bracts and sepals. Leaf-lamina very variable in shape and size, usually ovate, elliptic or oblong to lanceolate, 1–11 × 0·3–4·5 cm., obtuse or acute at the apex, truncate, rounded or even very slightly hastate, emarginate or subcordate at the base, covered on both surfaces with long fulvous hairs, rarely glabrous; petiole 0·5–2·5 cm. long. Flowers solitary or in small heads; peduncle 2–7 cm. long; bracts hairy at apex or middle of peduncle, linear to ovate, foliaceous, 0·3–2 cm.; pedicels somewhat thickened, 0·4–2 cm. long. Sepals unequal, ovate, acute or drawn out to a narrow point at the apex, rounded or subcordate at the base, the outer 1–2·5 × 0·5–1 cm., the inner narrower 0·6–2·2 × 0·2–0·8 cm. Corolla funnel-shaped, white, red or white with darker purple centre, 3–6·5 cm. long, midpetaline areas distinctly margined by raised veins, shortly hairy towards the apex. Ovary glabrous. Capsule globose, glabrous, apiculate. Seeds glabrous or nearly so.

Var. **crassipes** Verdc. in Kew Bull. **14**: 339 (1960); in F.T.E.A., Convolvulaceae: 98 (1963). Type as for the species.

    Ipomoea calystegioides E. Mey. ex Drege, Zwei Pflanz. Geogr. Doc.: 145, 153 (1843) *nomen nudum*.

    Aniseia calystegioides Choisy in DC., Prodr. **9**: 131 (1845). Type from S. Africa.

    Ipomoea calystegioides Choisy Hall. f. in Engl. Bot. Jahrb. **18**: 127 (1893).—Rendle in Journ. Linn. Soc., Bot. **40**: 150 (1911). Type from S. Africa.

    Ipomoea ukambensis Vatke in Linnaea **43**: 510 (1882). Type from Kenya.

    Ipomoea adumbrata Rendle & Britten in Journ. Bot. **32**: 173 (1894).—Baker & Rendle in Dyer, F.T.A. **4**: 145 (1905). Type from Angola.

    Ipomoea oblongata Dammer in Engl., Pflanzenw. Ost.-Afr. **C**: 331 (1895) *nom. illegit.* Type from Tanzania.

Ipomoea greenstockii Rendle in Journ. Bot. **34**: 38 (1896 et **39**: 14 (1901).—Baker & Wright in Dyer, F. C. **4**: 51 (1904). Type from S. Africa (Transvaal).

Ipomoea crassipes Hook. var. *ukambensis* (Vatke) Hall. f. in Bull. Herb. Boiss. **7**: 46 (1899).—Baker & Rendle in F.T.A. **4**, 2: 147 (1905). Type as for *Ipomoea ukambensis*.

Ipomoea sarmentacea Rendle in Journ. Bot. **39**: 15 (1901).—Baker & Wright in Fl. Cap. **4**: 57 (1904). Type from S. Africa (Transvaal).

Ipomoea bellecomans Rendle in Journ. Bot. **39**: 15 (1901).—Baker & Wright in Fl. Cap. **4**: 55 (1904). Type from S. Africa (Transvaal).

Ipomoea crassipes var. *shirensis* Baker in F.T.A. **4**, 2: 147 (1905). Type: Malawi, Shire Highlands, *Buchana* 89 (K, holotype).

Botswana. N: c. 19 km. S. of Botswana border, fl. & fr. 14.i.1960, *Leach & Noel* 23 (SRGH). SE: Mahalapye, St. Patrick's Mission, fl. & fr. 8.iii.1977, *Camerik* 87 (K; PRE). **Zambia**. N: Isoka, fl. & fr. 21.xii.1962, *Fanshawe* 7199 (K; SRGH). W: Mwinilunga, just N. of Mwinilunga, fl. & fr. 26.i.1938, *Milne-Redhead* 4357 (BM; K). C: Kapiri Mposhi, fl. & fr. 14.x.1964, *Fanshawe* 8949 (K; SRGH). S: Namwala, Rhino Pools, Kafue National Park, fl. & fr. 11.xii.1962, *Mitchell* 15/93 (COI; SRGH). **Zimbabwe**. N: Urungwe, Zwipani, 1070 m., fl. & fr. 12.x.1957, *Phipps* 777 (COI; SRGH). W: Bulalima Mangwe, about 1·5 km. W. of Plumtree, 1300 m., fl. & fr. 30.xi.1972, *Norrgrann* 292 (K; SRGH). C: Umvuma, fl. & fr. i.1923, *Eyles* 7556 (K; SRGH). E: Chipinge, 6 km. S. of Rusongo, 400 m., fl. & fr. 1.ii.1975, *Biegel, Pope & Gibbs-Russell* 4906 (K; SRGH). S: 48–97 km. S. of Masvingo, fl. & fr. 16.iv.1948, *Rodin* 4243 (K; PRE; SRGH). **Malawi**. C: About 3 km. N. of Kasungu, 1000 m., fl. & fr. 14.i.1959, *Robson* 1170 (K; LISC; SRGH). S: Makwapala, 716 m., fl. 3.iv.1937, *Lawrence* 354 (K). **Mozambique**. N: Montepuez, between Montepuez and Nantulo, 500 m., fl. & fr. 27.xii.1963, *Torre & Paiva* 9744 (BR; EA; K; LISC; PRE). Z: Namagoa, 200 km. inland from Quelimane, 30–60 m., fl. & fr. ix.1943, *Faulkner* 85 (BM; K; PRE). T: Angonia, fl. & fr. 19.xi.1980, *Macuácua* 1254 (LMA). MS: Manica, from Messambuzi to Mavita, 880 m., fl. 2.xi.1965, *Correia* 307 (LISC). GI: Canicado, Lower Mossingir, 305 m., fl. & fr. 1.xi.1905, *Swynnerton* 312 (BM). M: Maputo, nr. Goba, fl. & fr. 23.xi.1942, *Mendonça* 3085 (C; LISC; LMU).

Also in Kenya, Tanzania, S. Africa (Transvaal, Natal), Swaziland, Angola and Namibia. Woodland, savanna, grassland and sandy soil; 120–1400 m.

The var. *hewittioides* has prostrate stems. The leaf lamina is oblong-ovate and the corolla 4–5 cm. long. It occurs in Tanzania, Angola and possibly S. Africa.

16. **Ipomoea sinensis** (Desr.) Choisy in Mem. Soc. Phys. Genève **6**: 459 (1834); in DC., Prodr. **2**: 370 (1845).—Hemsely in Journ. Linn. Soc. Bot. **26**: 162 (1890).—Meeuse in Bothalia **6**: 728 (1958).—Verdc. in Kew Bull. **13**: 204 (1958); in F.T.E.A., Convolvulaceae: 100 (1963).—Heine in F.W.T.A. ed. 2, **2**: (1963).—Roessler in Merxm. Prodr. Fl. SW. Afr. **116**: 16 349 (1967).—Ross, Fl. Natal: 296 (1972).—Compton, Fl. Swaziland: 478 (1976).—Gibbs-Russell in Kirkia **10**: 490 (1977). TAB. **18**. Type from China.

Annual herb. Stems numerous from the apex of a taproot, prostrate or twining, or young shoots occasionally suberect, up to 2·5 m., more or less densely covered with white spreading hairs at least when young. Leaf lamina ovate to oblong ovate, rarely, rhombic, 2–10 × 2·5–7·5 cm., obtuse at the apex, broadly cordate or sometimes subhastate at the base, pubescent to glabrous; petiole slender, 1–10 cm. long, pilose. Inflorescence cymosely 1–3-flowered; peduncle slender, up to 6·5 cm. long, hairy; pedicels up to 1·9 cm. long, pilose, at first erect, reflexed in fruit; bracts minute, lanceolate. Sepals accrescent, 5–23 × 3–9 mm., glabrous to pilose, generally ciliate, unequal: outer ones ovate with cordate base or with broad, ovate or triangular, cordate or subhastate base 2–5 × 2–6 mm. and a narrow lanceolate 2–10 × 1·5–10 mm. acute apex; inner ones lanceolate, broadened at the base, 6·5 × 2 mm. Corolla funnel-shaped, pale mauve, white with purple centre, pink with greenish stripes or pure white, 1·2–2·2 cm.; midpetaline areas strigose (at least outside towards the apices). Capsule globose, glabrous, apiculate by the persistent style-base. Seeds velvety pubescent with grey or fawn hairs up to 5 mm. long.

Sepals (in flower) 4–11 × 4–8 mm., cordate at the base; sepals (in fruit) up to 23 × 9 mm.; corolla
usually white, up to 2·2 cm. long -   ·   ·   ·   ·   ·   ·   ·   subsp. *sinensis*
Sepals (in flower) with base about 2 × 2–3 mm. and apical part 4 × 1 mm.; sepals (in fruit) about
5 × 5–6 more or less 2–3 × 1·5 mm.; corolla usually pink or mauve with dark centre, about
1·3 cm. long -   ·   ·   ·   ·   ·   ·   ·   ·   subsp. *blepharosepala*

Subsp. **sinensis**.—Meeuse in Bothalia **6**: 729 (1958).—Verdc. in Kew Bull. **13**: 204 (1958); in F.T.E.A, Convolvulaceae: 100 (1963). Type as for the species.

Convolvulus sinensis Desr. in Lam., Encycl. Meth. Bot. **3**: 557 (1791). Type as above.
Convolvulus calycinus Roxb., Fl. Ind., ed. Carey & Wall. **2**: 51 (1824). Type from India.

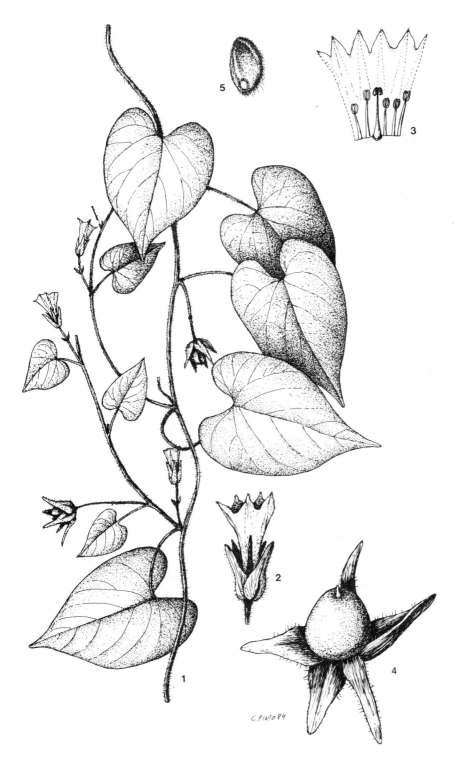

C.PINTO 84

Tab. 18. ɪᴘᴏᴍᴏᴇᴀ sɪɴᴇɴsɪs. 1, habit (×½); 2, flower (×1¼); 3, corolla opened to show stamens and pistil (×1¼); 4, fruit (×2); 5, seed (×2). All from *Lemos & Macuácua* 116.

Convolvulus ser Spreng., Syst. Veg. **1**: 598 (1824) *nom. illegit.* Type as for *Convolvulus sinensis*.

Convolvulus hardwickii Spreng., Syst. Veg. **4**, 2: 60 (1827) *nom. illegit.* Type as for *Convolvulus calycinus*.

Ipomoea hardwickii (Spreng.) Sweet, Hort. Brit., ed. 2: 372 (1830).—Hemsl. in Journ. Linn. Soc., Bot. **26**: 160 (1890). Type as above.

Aniseia calycina (Roxb.) Choisy in Mém. Soc. Phys. Genève **6**: 482 (1834); in DC., Prodr. **9**: 429 (1845). Type as for *Convolvulus calycinus*.

Ipomoea calycina (Roxb.) C. B. Clarke in Hook. f., Fl. Brit. India **4**: 201 (1883) non Meisn. (1869). Type as above.

Ipomoea auxocalyx Pilger in Notizbl. Bot. Gart. Berl. **11**: 819 (1933). Type from Tanzania.

Botswana. N: Chobe-Zambezi confluence, fl. & fr. 11.iv.1955, *Exell, Mendonça & Wild* 1467 (BM; LISC; SRGH). **Zambia**. E: Lundazi, Lukusuzi National Park, 800 m., fl. & fr. 12.iv.1971, *Sayer* 1165 (K; SRGH). S: Nr. Vwa's Village, Gwembe Valley, Zongwe R., fl. & fr. 1.iv.1952, *White* 2386 (FHO; K). **Zimbabwe**. N: Kariba, 457 m., fl. & fr. vi.1960, *Goldsmith* 74/60 (K; SRGH). W: Hwange, Deka R., fl. & fr. 21.vi.1934, *Eyles* 7957 (BM; K; SRGH). S: Buhera, 30 km. N. of Birchenough Bridge, 525 m., fl. & fr. 23.iv.1969, *Biegel* 2967 (K; LISC; SRGH). **Mozambique**. N., 3 km. from Montepuez to Nantulo, 500 m., fl. & fr. iv.1964, *Torre & Paiva* 11714 (C; LISC; LMA; WAG). T: Tete, Boroma, Msusa, fl. & fr. 25.vii.1950, *Chase* 2798 (BM; CUI; K; LISC; SRGH). MS: Chemba, Chiou, Exp. Station CICA, fl. & fr. 18.iv.1960, *Lemos & Macuácua* 116 (BM; COI; K; LISC; LMA; SRGH). GI: Gaza, Chicualacuala, fl. & fr. 27.iii.1974, *Balsinhas & Santos* 2660 (LMA). M: Maputo, Matola, Movene R., fl. & fr. 8.iv.1967, *Marques* 1960 (LISC).

Also in Tanzania, Swaziland, Namibia, through tropical Asia to Formosa. Woodland, thicket, grassland, aluvial and sandy soil; 450–900 m.

Subsp. **blepharosepala** (Hochst. ex A. Rich.) Verdc. ex Meeuse in Bothalia **6**: 729 (1958).—Verdc. in F.T.E.A., Convolvulaceae: 459 (1963).—Roessler in Merxm. Prodr. Fl. SW. Afr. **116**: 16 (1967).—Ross, Fl. Natal: 296 (1972). Type from Ethiopia.

Ipomoea cardiosepala Hochst. ex Choisy in DC., Prodr., **9**: 393, 429 (1845) *nom. nud.*

Ipomoea blepharosepala Hochst. ex A. Rich., Tent. Fl. Abyss. **2**: 72 (1851). Type from Ethiopia.

Ipomoea hardwickii sensu auctt. afr. non (Spreng) Sweet (1890).

Ipomoea cardiosepala Hochst. ex Baker & Wright in Dyer, F.C. **4**: 61 (1904).—Baker & Rendle in F.T.A. **4**: 2: 147 (1905) non Meisn. (1869).—N.E. Br. in Kew Bull. **1909**: 122 (1909).—Eyles in Trans. Roy. Soc. Afr. **5**: 453 (1916). Type from S. Africa.

All the names applied to subsp. *sinensis* have also been wrongly applied here.

Botswana. N: Maun, fl. & fr. 19.iii.1975, *Smith* 1289 (K; SRGH). SW: Ghanzi, Hide Store, 945 m., fl. & fr. 27.i.1970, *Brown* 8258 (K; SRGH). SE: Orapa, Airport Rd., fl. & fr. 19.iii.1974, *Allen* 33 (PRE; SRGH). **Zambia**. B: Masese, fl. & fr. 10.v.1961, *Fanshawe* (K; SRGH). **Zimbabwe**. N: Gokwe, Senanga Research Station, fl. & fr. 12.iv.1969, *Jacobsen* 581 (SRGH). C: Gweru, fl. & fr. v. 1925, *Earthy* 26189 (PRE). W: Hwange National Park, Makwa Pan, fl. & fr. 17.iv.1972, *Grosvenor* 693 (K; LISC; SRGH). E: Chipinge, Sabi Valley, fl. & fr. × x.1959, *Soane* 113 (K; LISC; SRGH). S: Gwanda, nr. Chiturupadzi Store, c. 40 km. NNE. of the Bubye-Limpopo confluence, fl. & fr. 12.v.1958, *Drummond* 5766 (K; SRGH). **Malawi**. S: Machinga, Liwonde National Park, Chigume Hill, fl. & fr. 17.iv.1980, *Blackmore, Brummitt & Banda* 1268 (K). **Mozambique**. T: Tete, 1 km., from Changara to Tete, 200 m., fl. & fr. 10.v.1971, *Torre & Correia* 18356 (LISC; LMA; LMU; WAG). GI: Caniçado, 4 km. from Caniçado on Souzuanine Rd., fl. & fr. 11.vi.1960, *Lemos & Balsinhas* 77 (BM; COI; K; LISC; LMA; SRGH). M: Sabie, Moamba, 1.v.1953, *Myre & Balsinhas* 1651 (LMA).

Ethiopia, Somalia, Socotra, Uganda, Kenya, Tanzania, Angola, Namibia, S. Africa (Swaziland and Natal) and also Arabia. Woodland, Acacia scrub, cultivated land, sandy and calcareous soils and roadsides; 200–1600 m.

17. **Ipomoea tenuirostris** Steud. ex Choisy in DC., Prodr. **9**: 379 (1845).—A. Rich., Tent. Fl. Abyss. **2**: 70 (1851).—Engl., Hochgebirgsfl. Trop. Afr.: 346 (1892); in Ann. Bot. Instit. Roma **7**: 229 (1892).—Hall. f. in Engl., Bot. Jahrb. **18**: 128 (1893).—Baker & Rendle in F.T.A. **4**: 143 (1905).—Brenan in Mem. N.Y. Bot. Gard. **9**: 7 (1954).—Dandy in F. W. Andr., Fl. Pl. Anglo-Egypt. Sudan **3**: 115 (1956).—Verdc. in Kew Bull. **13**: 202 (1958); in F.T.E.A., Convolvulaceae: 101, fig. 22, 1 and 10 (1963).—Heine in F.W.T.A. ed. 2, **2**: 349 (1963).—Binns, H.C.L.M.: 40 (1968). Type from Ethiopia.

Perennial herb. Stems slender, twining or prostrate, up to 3 m., more or less densely covered with yellowish spreading hairs. Leaf lamina ovate to oblong, 4–12 × 3·5–9·5 cm., acute to rounded at the mucronate apex, cordate at the base, entire, pubescent or hairy on both surfaces; petiole 2·5–8·5 cm. long, hairy. Flowers in 2–5-flowered often lax, sometimes dense cymes; peduncle 1·5–10·5 cm. long pedicels 1–3·5 cm. long; bracts lanceolate, more or less 3 mm. long, hairy like the sepals. Sepals ovate to lanceolate, 7–12 × 1·5–3 mm., narrowed at the apex, yellowish hairy, minutely hirsute on the reverse and margin; inner ones narrower. Corolla funnel-shaped, white to mauve, usually, 2–4 cm. long, midpetaline areas very distinctly purple, at least at the centre minutely hirsute. Capsule globose, glabrous, tipped with indurated persistent style base. Seeds brown, minutely velvety sometimes with long white hairs on the upper angles.

Subsp. **tenuirostris.**—Verdc. in Kew Bull. **13**: 203 (1958); in F.T.E.A., Convolvulaceae: 101 (1963). Type as for the species.

  Ipomoea acuminata Baker in Kew Bull. **1894**: 72 (1894) nom. illegit. (Vahl) Roem. & Schultes (1819). Type: Malawi, "Zambesi Highlands", *Kirk* (K, lectotype).

  Ipomoea zambesiaca Britten in Journ. Bot. **32**: 85 (Mar. 1894) *nom. illegit.* non Baker (Feb. 1894). Type as for *Ipomoea acuminata* Baker.

  Ipomoea halleriana Britten in Journ. Bot. **32**: 170 (1894). Type as for *Ipomoea acuminata* Baker.

  Ipomoea mweroensis Baker in Kew Bull. **1895**: 291. Type: Zambia, Lake Mweru, *Carson* 23 (K, holotype).

  Ipomoea gracilior Rendle in Journ. Bot. **46**: 180 (1908). Type from Uganda.

Flowering on long twining stems. Leaf-blade ovate to oblong, 4–12 × 3·5–9 cm; acute, mostly more or less densely pilose. Inflorescences mostly several-flowered. Sepals pilose. Corolla usually purple, 2–4 cm long.

Zambia. N: Kawambwa, fl. & fr. 7.viii.1958, *Fanshawe* 4670 (K). W: Kitwe, fl. & fr. 6.viii.1969, *Mutimushi* 3516 (K; SRGH). **Zimbabwe.** W: Bulawayo, Bellevue, fl. & fr. ix.1908, *Chubb* 335 (BM). E: Inyanga, Pungwe Valley, fl. & fr. 17.vii.1948, *Chase* 899 (BM; K; SRGH). **Malawi.** N: Mzimba, Mzuzu, Marymount, fl. & fr. 1372 m., 27.vii.1972, *Pawek* 5539 (K; SRGH). C: Ntchisi Mt., 1600 m., fl. & fr. 26.vii.1946, *Brass* 16957 (BM; K; SRGH). S: Thyolo Mt., 1200 m., fl. & fr. 19.ix.1946, *Brass* 17649 (BM; K; SRGH). **Mozambique.** Z: Milange, Chiperone Mt., 900 m., fl. & fr. 8.ii.1972, *Correia & Marques* 2551 (LISC). T: Cahobra Bassa, near Meroeira, 625 m., fl. & fr. 3.ii.1973, *Torre, Carvalho & Ladeira* 18987 (C; LISC; LMA; MO; WAG). MS: Gorongosa Mt., 1300 m., fl. & fr. 10.vii.1969, *Leach & Cannell* 14273 (SRGH).

  Also in Ethiopia, Sudan, Congo Republic, Uganda, Kenya, Tanzania and Cameroun. Woodland, Grassland and cultivated ground; 600–1900 m.

  Subsp. *hindeana* (Rendle) Verdc. occurs in Kenya and Tanzania; Subsp. *repens* in Ethiopia and Kenya.

18. **Ipomoea linosepala** Hall. f. in Engl., Bot. Jahrb. **18**: 130 (1893).—Hiern, Cat. Afr. Pl. Welw. **1**: 734 (1898).—Baker & Rendle in F.T.A. **4**, 2: 150 (1905).—Rendle in Journ. Bot. **46**: 181 (1908). Type from Angola.

  Ipomoea xiphosepala Baker in Kew Bull. **1894**: 69 (1894). Type from Angola.

An erect perennial much-branched herb with a thick fusiform woody root. Stems branched from the base; branches up to 1·5 m. long, densely covered with long spreading yellowish hairs like the leaves. Leaf lamina ovate, becoming more or less oblanceolate to obovate, more or less 1·5–5 × 3–2·2 cm., shortly acute to obtuse at the apex, obtuse to subacute at the base; petiole up to 5 mm. long, densely hairy. Flowers generally solitary from the axils of the upper leaves; peduncle 5–15 mm. long, densely hairy; bracts linear-subulate up to 14 mm. long, below the flower densely hairy. Sepals linear-lanceolate below with a long linear subulate upper portion, up to 18 mm. long, densely ciliate like the young shoots. Corolla funnel-shaped-campanulate, pink to cream whitish with purple centre, 2–3·5 cm. long; midpetaline areas hairy in the upper portion. Capsule subspherical. Seeds blackish, minutely pubescent.

Zambia. N: Luapula Distr., Mbereshi, 1050 m., fl. & fr. 16.i.1960, *Richards* 12393 (K; SRGH). W: Mwinilunga, 1300 m., fl. & fr. 24.i.1975, *Brummitt, Chisumpa & Polhill* 14051 (K; LISC; NDO). **Malawi.** N: Chitipa Distr., just outside gate of Nyika National Park on Nthalire Rd., 1480 m., fl. & fr. 1.iii.1982, *Brummitt, Polhill & Banda* 16197 (K). C: Ntchisi Mt., 1450 m., fl. & fr. 20.ii.1959, *Robson* 1684 (BM; K; LISC; SRGH).

  Also in Angola. Woodland and mountain forest; 1280–1550 m.

19. **Ipomoea alpina** Rendle in Journ. Bot. **50**: 253 (1912).—Verdc. in Kew Bull. **13**: 205

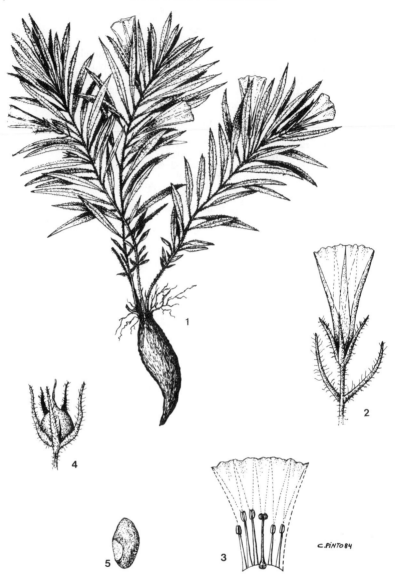

Tab. 19. IPOMOEA ALPINA. 1, habit ($\times\frac{1}{2}$), from *Robinson* 6028; 2, flower with bracts ($\times$1); 3, corolla opened to show stamens and pistil ($\times$1), 2–3 from *Brummitt, Polhill & Banda* 16235; 4, fruit ($\times$2); 5, seed ($\times$4), 4–5 from *Kornas* 3177.

(1958); in F.T.E.A., Convolvulaceae: 104 (1963).—Binns, H.C.L.M.: 40 (1968). TAB. **19** Type from Zaire.
    Ipomoea hockii De Wild. in Fedde, Repert. **11**: 539 (1913). Type from Zaire.

Perennial herb, bristly-pilose with orange-brown hairs, up to 1·7 m. high. Stems several, annual, erect or semiprostrate, up to 25 cm. tall, from a tuberous rootstock. Leaf lamina linear to oblong, 1·7–9 $\times$0·15–1·18 cm., more or less acute at the apex, cuneate at the base, sparsely bristly-pilose to densely pilose on both surfaces particularly on the margins and nerves beneath; petiole 2–5 mm. long. Flowers solitary in the upper axils; peduncle and pedicel together 0·3–3 cm. long; bracts linear, 0·6–2 cm. long. Sepals linear triangular, bristly pilose, 0·75–1·7 cm. long. Corolla tubular, little expanded above, orange-yellow, pale yellow or white, 1·8–3·7 cm. long, glabrous below but apices of midpetaline areas hairy; interpetaline areas tenuously thin. Capsule globose, glabrous. Seeds minutely pubescent.

Zambia. N: Kwambwa, Ntimbachushi Falls, c. 13 km. E. of Kawambwa, 1220 m., fl. & fr. 22.xii.1967, *Simon & Williamson* 1498 (K; LISC; SRGH). W: Kitwe, fl. & fr. 10.xii.1969, *Fanshawe* 10688 (SRGH). C: Serenje Distr., Kundalila Falls, 1500 m., fl. & fr. 4.ii.1973, *Strid* 2843 (K). **Malawi**. N: Nyika Plateau, Chelinda Bridge, fl. & fr. 10.i.1964, *Hilliard & Burtt* 4391 (K; SRGH). C: Dedza, Chongoni Forest, fl. & fr. 7.ii.1967, *Salubeni* 5546 (K; LISC; SRGH). S: Mulanje Mt., Chambe Plateau, 1830 m., fl. & fr. 14. II. 1979, *Blackmore, Brummitt & Banda* 425 (K). **Mozambique**. T: Masanga, Furancungo Mt., 1380–1420 m., *Pereira, Sarmento & Marques* 1734 (LMA).

Also in Tanzania and Shaba Province of Zaire. Woodland, grassland and sandy soil; 1050–2750 m.

20. **Ipomoea polhillii** Verdc. in Kew Bull. **21**: 89 (1967). Type from Tanzania.

Perennial herb with all parts more or less densely covered with long spreading and appressed hairs. Stems prostrate, several radiating creeping shoots from a narrow rootstock, slender, more or less densely covered with long spreading and appressed yellowish hairs. Leaf lamina oblong-ovate, 1·5–4 × 1–2 cm., entire, acute or more or less rounded at the mucronulate apex, rounded to subcordate at the base, more or less densely covered with appressed yellow and whitish hairs; petiole 3–8 mm. long, hairy as the stems. Inflorescences axillary, 1–2 flowered; peduncle 2–8 mm. long; pedicels 3–20 mm. long; bracts linear up to 3 mm. long. Sepals lanceolate 6·5–7 mm. long, acuminate, the inner ones with hyaline margin, all covered with spreading yellowish hairs. Corolla funnel-shaped, white, more or less 2–7 cm. long, midpetaline areas fuscous and hairy at the apices. Capsule and seeds unknown.

Malawi. N: Mzimba Distr., Viphya Plateau, E. of Champira, 1676 m., fl. 11.i.1975, *Pawek* 8920 (K; SRGH).

Also in Tanzania. On rocky soil; 1676 m.

21. **Ipomoea involucrata** Beauv., Fl. Owar. **2**: 52, tab. 89 (1817).—Choisy in DC., Prodr. **9**: 365 (1845).—Hall. f. in Engl., Bot. Jahrb. **18**: 135 (1893), ex parte.—Dammer in Pflanzenw. Ost-Afr. **C**: 332 (1895).—Hall. f. in Bull. Herb. Boiss. **5**: 375 (1897).—Hiern. Cat. Afr. Pl. Welw. **1**: 735 (1898).—Baker & Rendle in F.T.A. **4**, 2: 150 (1905).—Eyles in Trans. Roy. Soc. S. Afr. **5**: 454 (1916).—Fries, Wiss. Ergebn. Schwed. Rhod.-Kongo-Exped. **1**: 269 (1916).—Brenan in Mem. New York Bot. Gard. **9**: 7 (1954). Dandy in F. W. Andr., Fl. Pl. Anglo-Egypt. Sudan **3**: 112 (1956).—Meeuse in Bothalia **6**: 744 (1958). —Verdc. in Kew Bull. **13**: 206 (1958); in F.T.E.A., Convolvulaceae: 104 (1963).—Heine in F.W.T.A. ed 2, **2**: 347, fig. 283 (1963).—Binns, H.C.L.M.: 40 (1968). Type from Nigeria.

An extremely variable annual or perennial herb. Stems slender, twining, finely and usually retrorsely hairy to glabrescent, up to 4 m. long. Leaf lamina ovate-cordate up to 13·5 × 13 cm., acute or acuminate at the apex, hairy to villous on both surfaces, sometimes more densely beneath; petiole slender, 1·3–8 cm. long, retrorsely pilose like the stems and peduncles. Inflorescence a dense head, enclosed in a large foliaceous boat-shaped involucre; peduncle 1–16 cm. long; flowers few to many, very shortly pedicellate; outer bracts connate in a hairy boat-shaped structure, 3–6 cm. in diam. with 2 cusps, more or less pubescent; inner bracts small, obovate to linear-oblong, acute to aristate. Sepals 6–15 × 1·5–4 mm., glabrescent to hairy, margins setose; the outer ones lanceolate; the inner ones shorter and more ovate. Corolla funnel-shaped, purple, rose, white or white with pink throat up to 5·5 cm. long, 2–5 cm. wide at the mouth, about 4–10 mm. wide at the base, sparsely pilose on the midpetaline areas. Capsule globose, glabrous. Seeds blackish, glabrous or shortly pubescent.

1. Perennial, densely villous or pilose; involucre often incompletely connate; outer bracts one below the other often 3-pointed instead of 2, very acuminate at the apices; inflorescence usually profusely many-flowered, the flowers obscuring the inner bracts; corolla about 3·5 cm. long - - - - - - - - - - - var. *operosa*
— Perennial or annual, more or less velvety pubescent or bristly pubescent but not densely

villous; outer bracts completely joined forming a continuous involucre; inflorescence usually
few-flowered · · · · · · · · · · · · · · · · 2
2. Perennial or annual; involucre up to 10 cm. long; sepals typically ovate-lanceolate, about
1·5 × 0·4 cm., pilose; corolla 2·5–5 cm. long · · · · · var. *involucrata*
— Annual; involucre small, up to 4·5 cm. long, resembling that of *I. pileata*; sepals small,
lanceolate, 0·6 × 0·15 cm., not long pilose, often glabrescent, closely veined; corolla 2·3 cm.
long · · · · · · · · · · · · · · · var. *burttii*

Var. **involucrata**. Verdc. in Kew Bull. **13**: 206 (1958); in F.T.E.A., Convolvulaceae: 104
(1963).—Binns, H.C.L.M.: 40 (1968).
  Convolvulus perfoliatus Schumach & Thonn., Beskr. Guin. Pl.: 89 (1827). Type from
Ghana.

Zambia. N: Mbala, 1680 m., fl. & fr. 26.v.1952, *Siame* 65a (BM; LISC). W: Mwinilunga
Distr., Zambezi rapids, 6 km. N. of Kalene Hill, 1350 m., fl. & fr. 20.ii.1975, *Hooper & Townsend*
230 (K). C: nr. Chilanga, Mt., Makulu Research Stat., fl. & fr. 17.iii.1960, *Angus* 2156 (LISC;
SRGH). S: Between Pemba and Mazabuka, fl. & fr. 11.vii.1930, *Pole Evans* 7804 (PRE).
Zimbabwe. N: Miami, fl. & fr. v.1926, *Rand* 120 (BM). C: Harare, fl. & fr. vii.1898, *Rand* 561
(BM). E: Mutare Distr., Norseland, Vumba 1615, fl. & fr. 21.iv.1950, *Chase* 2163 (BM; K;
LISC; SRGH). S: Masvingo Distr., fl. & fr. xii.1909, *Monro* 1268 (BM). **Malawi**. N: Rumphi
Distr., Nyika Plateau, 12 km., from Chelinda Camp on Rd. to Katumbi, 2260 m., fl. & fr.
17.v.1970, *Brummitt* 10843 (K; LISC; MAL; PRE; SRGH). C: Lilongwe, 1100 m., fl. &
fr. 29.iii.1970, *Brummitt & Little* 9496 (K; LISC; MAL; PRE; SRGH). S: Limbe, 1220 m., fl. &
fr. v.1948, *Goodwin* 124 (BM). **Mozambique**. N: Ribáuè, S. face of Ribáuè Mt., 975 m., fl. & fr.
19.vii.1962, *Leach & Schelpe* 11404 (K; LISC; SRGH). T: Cahobra Bassa, between 700 and 818
m., fl. & fr. 4.v.1972, *Pereira & Correia* 2384D (LISC; LMU). MS: Gorongosa, Gogogo Mt.,
1770 m., fl. & fr. 5.vii.1955, *Schelpe* 456 (LISC; SRGH).
Throughout tropical Africa from West Africa to Angola and S. Africa (Transvaal). Secondary
woodland, open and marginal forest, bushland, grassland, stream beds, weed of cultivations,
sandy soil and roadside; 700–2300 m.

Var. **operosa** (C. H. Wright) Hall. f. ex Verdc. in Kew Bull. **13**: 206 (1958); in F.T.E.A.,
Convolvulaceae: 105 (1963).—Binns, H.C.L.M.: 40 (1968). Type: Malawi, Zomba, *Kirk*,
Whyte (K, syntypes).
  Ipomoea operosa C. H. Wright in Kew Bull. 1897.—Brenan in Mem. N. Y. Bot. Gard. 9: 7
(1954). Type as above.

Zambia. N: Mbala Distr., Old Kasama Rd., 1525 m., 4.iv.1952, *Richards* 1343 (K). **Malawi**.
N: Mzimba Distr., Viphya Plateau, 5 km. S. of Chikangawa, 1720 m., 8.v.1970, *Brummitt* 10446
(K; SRGH). S: Mulanje, Nampende/Migowi spur, Machese Mt., 1830 m., 25.vii.1958, *Chapman* 729 (K; LISC; SRGH). **Mozambique**. N: Nampula, 1947, *Andrada* 758 (LISU). Z:
Gúruè Mt., 1.x.1941, *Torre* (C; LISC; MÒ; UPS).
Also in Tanzania. Evergreen forest, bushland, grassland; 1500–2200 m.

Var. **burttii** Verdc. in Kew Bull. **13**: 206 (1958); in E.T.E.A., Convolvulaceae: 105 (1963). Type
from Tanzania.
  Zambia. B: Masese, fl. & fr. 3.v.1961, *Fanshawe* 6538 (K). N: Mbala, 1525 m., fl. & fr.
29.ii.1952, *Richards* 849 (K). W: Kabompo on Zambezi Rd., 24.iii.1961, *Drummond &
Rutherford-Smith* 7274 (K; SRGH). S: Bombwe Forest, fl. & fr. 1934, *Martin* 628 (K).
Also in Tanzania. Woodland, damp and sandy soils; 1200–1550 m.

22. **Ipomoea pileata** Roxb., Fl. Ind., ed. Carey & Wall. **2**: 94 (1824), ed. Carey 1: 504
(1832).—Choisy in DC., Prodr. **9**: 365 (1845).— Baker & Wright in Dyer, F.C. **4**: 53
(1904).—Baker & Rendle in F.T.A. **4**, 2: 151 (1905).—Eyles in Trans. Roy. Soc. S. Afr. **5**:
455 (1916).—van Oosstr. in Blumea **3**: 507 (1940); in Fl. Males., Ser. 1, **4**: 467
(1953).—Dandy in F. W. Andr., Fl. Pl. Anglo-Egypt. Sudan **3**: 112 (1956).—Meeuse in
Bothalia **6**: 745 (1958).—Verdc. in F.T.E.A., Convolvulaceae: 105 (1963).—Heine in
F.W.T.A. ed. 2, **2**: 347 (1963).—Binns, H.C.L.M.: 40 (1968). TAB. **20**. Type a plant
cultivated at Calcutta from Chinese seed.
  Ipomoea involucrata sensu Hall. f. in Engl., Bot. Jahrb. **18**: 135 (1893) ex parte, excl.
type.
  Ipomoea involucrata Beauv. var. *albiflora* Hiern in Cat. Afr. Pl. Welw. **1**: 735 (1898).
Type from Angola.
  Ipomoea involucrata sensu Rendle in F.T.A. **4**, 2: 150 (1905) pro parte, quoad specim. ex
Kisantu, Zaire, non Beauv.

Annual or sometimes perennial with a fusiform thick taproot, very similar to *I.
involucrata*. Stems twining, pubescent up to 3 m. long. Leaf lamina ovate, up to 12·5 ×
10 cm., acute or acuminate, cordate, pubescent; petiole up to 15 cm. long. Inflorescence a pedunculate involucrate head, few to many-flowered, enclosed in a 2–5·5(7)

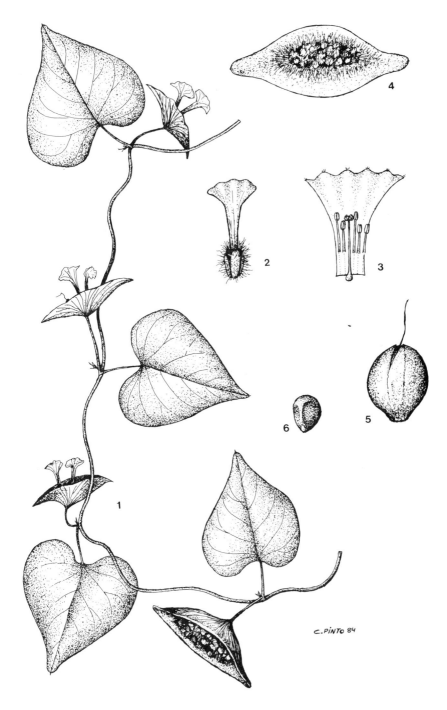

Tab. 20. ɪᴘᴏᴍᴏᴇᴀ ᴘɪʟᴇᴀᴛᴀ. 1, habit (×½); 2, flower (×1); 3, corolla opened to show stamens and pistil (×1), from *Brummitt* 11027; 4, inflorescence, top view (×¾); 5, fruit (×2); 6, seed (×2), 4–6 from *Loveridge* 1838.

pubescent boat-shaped involucre, 2–5·5(7) cm. long, originating from the outer bracts, connate, with two cusps; peduncle retrorsely pilose, 1–7 cm. long; inner bracts much smaller, ovate, very obtuse, up to 1·7 cm. long, vilous at the base. Sepals 3: outer ones oblong-spathulate to oblong, obtuse, more or less 10 mm. long; inner ones much narrower, lanceolate with a long and slender point, more or less 9 mm. long, all hairy outside. Corolla funnel-shaped, pink or white with darker centre; tube up to 25 × 1·5–2 mm. at the base; limb about 10 × 15 mm., midpetaline areas sparsely pilose, mucronate. Capsule ovoid glabrous. Seeds ovoid, blackish, glabrous or shortly pubescent around the hilum and sometimes on the angles.

Zambia. B: Masese, fl. & fr. 13.iii.1961, *Fanshawe* 6416 (K; SRGH). N: Mbala Distr., Chilongowelo Drive, 1465 m., fl. & fr. 2.iv.1952, *Richards* 1380 (K). W: Ndola, fl. & fr. 20.iii.1954, *Fanshawe* 986 (K). C: Mt. Makulu Research Station, c. 16 km. S. of Lusaka, fl. & fr. 31.iii.1960, *Smartt* 5 (LISC; SRGH). E: Lundazi, Lukusuzi National Park, 800 m., fl. & fr. 12.iv.1971, *Sayer* 1175 (SRGH). S: Kalomo, Mulobezi, fl. & fr. 9.iv.1955, *Exell, Mendonça & Wild* 1439 (BM; LISC; SRGH). **Zimbabwe**. N: Gokwe, Sengwa Research Station, fl. & fr. 3.iv.1969, *Jacobsen* 539 (K; PRE; SRGH). W: Hwange, fl. & fr. iv.1932, *Levy* 33 (PRE). C: Harare, Beatrice, 1310 m., fl. & fr. 17.iii.1961, *Drewe* 59 (SRGH). S: Masvingo, Kyle National Park, fl. & fr. 16.iii.1976, *Basera* 246870 (SRGH). **Malawi**. N: Nkahata Bay Distr., c. 3·2 km. S. on Lakeshore Rd., 610 m., fl. & fr. 3.vi.1973, *Pawek* 6817 (K; SRGH). C: Lilongwe Distr., beside Ngala Mt., 23 km. SE. of Lilongwe, 1220 m., fl. & fr. 28.iv.1970, *Brummitt* 10245 (K; LISC; MAL SRGH). S: Manwera Escarpment, Jalasi, 1120 m., fl. & fr. 15.iii.1955, *Exell, Mendonça & Wild* 911 (BM; LISC; SRGH). **Mozambique**. N: Macondes, 1 km. from Chomba to Negomano, 800 m., fl. & fr. 13.iv.1964, *Torre & Paiva* 11899 (C: LISC). Z: Maganja da Costa, Bajone Gurai, fl. & fr. 19.v.1971, *Balsinhas* 1883 (LMA). T: Tete, between Songo and Barrage, fl. & fr. 25.iii.1972, *Macedo* 5091 (LISC; LMA; LMU). MS: Manica, southern tip Chimanimani Mts., nr. Lusitu R., 305 m., fl. & fr. 29.v.1969, *Müller* 1155 (K; LISC; SRGH). GI: between Chibuto and Manjacaze, 15 km. from Chibuto, fl. & fr. 10.v.1973, *Balsinhas* 2508 (LMA).

Also in Gambia, Zaire, Angola, S. Africa (Transvaal), Mascarene Islands; India to China and Malaysia? Woodland, edge of rain forest, deciduous thicket, grassland, river banks, weed of cultivated land, roadside, rocky and sandy soil; 300–1530 m.

23. **Ipomoea crepidiformis** Hall. f. in Engl., Bot. Jahrb. **18**: 131 (1893).—Dammer in Engl., Pflanzenw. Ost.-Afr. **C**: 332 (1895).—Baker & Rendle in F.T.A. **4**, 2: 152 (1905).—Verdc. in F.T.E.A., Convolvulaceae: 106 (1963). Type from Tanzania.

Annual or perennial herb. Stems erect or prostrate, up to 1 m. long, often reddish, appressed, strigose, with laxly leafy branches. Leaf lamina discolorous, oblong or linear-lanceolate, 2·7–8·5 × 0·25–2·7 cm., sparesely pilose above, densely to sparsely white-pilose beneath, at first white-silky; petiole up to 15 mm. long. Flowers congested in small densely pilose heads; peduncle up to 9 cm. long, rigidly erect, suberect or ascending; bracts small, setaceous, long ciliate. Sepals lanceolate, 6–10 mm. long, white-silky with a conspicuous green edge. Corolla funnel-shaped, red or mauve, 1·5–3·3 cm. long, the tube narrow below; midpetaline areas minutely silky outside. Capsule ovoid-globose, glabrous. Seeds pubescent.

1. Plant 6–15 cm. tall, (dwarf variety resembling the typical variety in all other respects)
                                                                                                   var. *minor*
— Plant 30–100 cm. tall ·   ·   ·   ·   ·   ·   ·   ·   ·   ·   ·   ·   2
2. Usually annual; stems not very leafy, erect or straggling, about 0·5 m. long; leaves more or less thinly white-pilose beneath; heads 13–16 × 8–10 mm. at the sepals
                                                                                          var. *crepidiformis*
— Perennial; stems leafy, erect or suberect, 0·30–1 m. long; leaves very densely white-pilose beneath; heads larger, 13 × 18–20 mm. at the sepals   ·   ·   ·   var. *microcephala*.

Var. **crepidiformis**. Verdc. in F.T.E.A., Convolvulaceae: 106 (1963).
    Ipomoea tanganyikensis Baker in Kew Bull. **1895**: 70 (1895). Type: Zambia, near Mbala, Fwambo, *Carson* 73 (K, holotype).

Zambia. N: Mporokoso, nr. Muzombwe, Mweru-Wa-Ntipa, 1040 m., fl. & fr. 15.iv.1961, *Phipps & Vessey-Fitzgerald* 3194 (K; LISC; SRGH). W: Luanshya, fl. & fr. 30.iii.1956, *Fanshawe* 2854 (K; SRGH). C: 62 km. W. of Lusaka on Mumbwa Rd., fl. & fr. 20.iii.1965, *Robinson* 6459 (K; SRGH). S: Mazabuka Distr., Choma-Namwala Rd., c. 19 km. N. of Choma, 1250 m., fl. & fr. 17.iii.1956, *Robinson* 1363 (K; SRGH). **Zimbabwe**. N: Binga, Sengwa Wildlife Research Area, fl. & fr. 29.iii.1974, *Hall-Martin* in GHS 231687 (SRGH). **Malawi**. C: Lilongwe Distr., Chitedze, 1150 m., fl. & fr. 22.iii.1955, *Exell, Mendonça & Wild* 1105 (BM; LISC; SRGH).

Also in Uganda, Kenya and Tanzania. Woodland, grassland, sandy soil and roadside; 600–1150 m.

Var. **minor**. Rendle in Journ. Bot. 39: 20 (1901).—Baker & Rendle in F.T.A. **4**, 2: 152 (1905).—Verdc. in F.T.E.A., Convolvulaceae: 166 (1963). Type from Kenya.
Ipomoea taborana Dammer in Engl. Pflanzenw. Ost.-Afr. **C**: 333 (1895). Type from Tanzania.

Zambia. N: Mbala Distr., Cassava Sands, Lake Tanganyika, 780 m., fl. & fr. 17.ii.1959, *Richards* 10941 (K; SRGH). **Zimbabwe**. C: Chegutu, Poole Farm, 1220 m., fl. & fr. 15.iv.1948, *Hornby* 2901 (K; SRGH).
Also in Kenya and Tanzania. Sand dunes; 780–1220 m.

Var. **microcephala**. (Hall. f.) Verdc. in Kew Bull. **13**: 207 (1958); in F.T.E.A., Convolvulaceae: 107 (1963). Type from Tanzania.
Ipomoea microcephala Hall. f. in Engl., Bot. Jahrb. **18**: 131 (1893).—Dammer in Engl., Pflanzenw. Ost-Afr. **C**: 332 (1895).—Baker & Rendle in F.T.A. **4**: 152 (1905).
Ipomoea kassneri Pilger in Engl., Bot, Jahrb. **48**: 350 (27 Aug. 1912) non Rendle (1 Aug. 1912). Type from Zaire.
Ipomoea falcata R. E. Fries in Notizbl. Bot. Gart. Berl. **10**: 99 (1927). Type from Kenya.

Zambia. N: Mbala Distr., Sunzu Mt., 1950 m., fl. & fr. 20.iv.1961, *Richards* 15081 (K; SRGH). W: Mufulira, fl. & fr. 3.i.1956, *Fanshawe* 2683 (K).
Also in Somalia, Kenya, Tanzania and Zaire. Woodland, grassland, sandy soil and roadside; 1500–1950 m.

24. **Ipomoea chloroneura** Hall. f. in Engl., Bot. Jahrb. **18**: 132 (1893).—Hiern, Cat. Afr. Pl. Welw. **1**: 734 (1898).—Hall. f. in Warb., Kunene-Samb.—Exped. Baum: 346 (1903). —Baker & Rendle in F.T.A. **4**, 2: 153 (1905).—N.E. Br. in Kew Bull. **1909**: 122 (1909).—Meeuse in Bothalia **6**: 739 (1958).—Verdc. in F.T.E.A., Convolvulaceae: 107 (1963).—Roessler in Merxm. Prodr. Fl. SW. Afr. **116**: 13 (1967). Type from Angola.

Annual up to 30 cm. tall. Main stem branched from the base, with branches ascending or prostrate up to 40 cm. long, often branched again, densely covered with appressed short white hairs and spreading golden-yellow hairs when young as are petioles and peduncles, less densely so when older. Leaf lamina entire, elliptic, oblong or lanceolate to oblanceolate or elliptic-obovate, 2·5–7 × 1·2–2·8 cm., obtuse or acute at the apex, cuneate at the base, appressed pilose above, densely silvery pilose beneath and with stiff yellowish hairs on the nerves; petiole 6–20 mm. long. Flowers few in small dense heads; peduncle 2·6–12 cm. long; outer bracts foliaceous up to 20 mm. long, resembling the young leaves; inner bracts smaller. Sepals subequal, elliptic-lanceolate, about 6 mm. long, aristate, margins glabrous below, apex hairy caudate, accrescent in fruit. Corolla funnel-shaped, white or pale yellow-green, about 1·5 cm. long, midpetaline areas hairy. Capsule globose, glabrous. Seeds silky-pubescent.

Caprivi Strip. E: Mpola., c. 24 km. from Katima Mulilo on Rd. to Ngoma, 915 m., fl. & fr. 5.i.1959, *Killick & Leistner* 3293 (SRGH). **Botswana**. N: 69 km. W. of Nokaneng, fl. & fr. 12.iii.1965, *Wild & Drummond* 6900 (K; LISC; SRGH). SE: Mochundi, ii.1914, *Rogers* 6557 (K). **Zambia**. B: Mongu Lealui, fl. & fr. 20.ii.1966, *Robinson* 6844A (SRGH). W: Kabompo, fl. & fr. 24.iii.1961, *Drummond & Rutherford-Smith* (7262 K; LISC; SRGH). S: Machili, fl. & fr. 22.ii.1961, *Fanshawe* 6308 (K). **Zimbabwe**. W: Hwange, fl. & fr. 17.ii.1956, *Wild* 4774 (COI; K; SRGH).
Also in Tanzania, Angola, S. Africa and Namibia. Woodland, savanna, grassland and sandy soil; 900–1100 m.

25. **Ipomoea ommaneyi** Rendle in Journ. Bot. **40**: 190 (1902) ("*ommanei*").—Baker & Wright in Dyer, F.C. **4**: 53 (1904).—Meeuse in Bothalia **6**: 740 (1958). Type from S. Africa (Transvaal).

Perennial with a thick fusiform tuberous taproot. Stems several, trailing, up to 2 m. long, terete, densely pubescent at least in the younger parts. Leaf lamina ovate-lanceolate or oblong-lanceolate, up to 13 × 7·5 cm., rounded to shallowly cordate at the base, subobtuse to acuminate at the apex, densely covered on both surfaces with silvery-white hairs, later somewhat glabrescent, densely yellow-ciliate on the margin, nerves often very prominent beneath and stout; petiole stout, 1–6 cm. long, terete, somewhat flattened and grooved above, densely hairy. Flowers several together in a more or less dense head; outer bracts ovate or ovate-subspathulate, acuminate-cuspidate up to 3 cm. long, densely silky as are the inner bracts and sepals; inner bracts shorter and narrower; peduncle up to 11 cm. long, densely hairy. Sepals 2–3·5 cm. long, the outer ones lanceolate, more or less acuminate-cuspidate, the inner ones much narrower, linear-lanceolate. Corolla funnel-shaped, rose-magenta, 5–7(9) cm. long; midpetaline areas white and densely silky outside. Ovary glabrous. Capsule globose, enclosed in the calyx. Seeds black glabrous.

Botswana. N: c. 109 km. from Maun on main Rd., Neyi Pan, 25.xii.1967, *Lambrecht* 475 (K; SRGH). SE: c. 3·2 km. E. of Kanye, 1340 m., fl. & fr. 18.i.1960, *Leach & Noel* 194 (SRGH). **Zambia.** E: S. of Lundazi, fl. & fr. 28.i.1969, *Anton-Smith* in GHS 200714 (SRGH). **Zimbabwe.** W: Matopos, Maleme Valley, fl. & fr. 10.i.1963, *Wild* 5986 (K; SRGH). C: Rusape, fl. & fr. iii.1953, *Dehn* 56 (K). E: Mutare, Odzani R., 11.ii.1948, *Chase* 981 (BM; SRGH). **Mozambique.** N: Ribáuè, 580, fl. & fr. 25.iv.1968, *Macedo* 3045 (LMA).
  Also in S. Africa. Pans, roadsides and rocky soil; 580–1350 m.

26. **Ipomoea atherstonei** Baker in Dyer, F.C. **4**: 53 (1904).—Baker & Rendle in F.T.A. **4**, 2: 154 (1905).—Meeuse in Bothalia **6**: 741 (1958).—Ross, Fl. Natal: 295 (1972). Type from S. Africa (Transvaal).

  Prostrate perennial with fusiform tuberous root. Stems stout, often angular up to 2 m. long, usually densely covered with short stiff hairs as are petioles, peduncles, bracts and pedicels. Leaf lamina oblong to ovate-elliptic, 6–14× 4–7·5 cm., broadly cuneate, truncate, rounded to shallowly cordate at the base, obtuse, emarginate or rounded and mucronate, occasionally acute, at the apex, thinly covered on both surfaces with appressed or somewhat spreading short stiff hairs, sometimes very densely so; petiole 1–7.5 cm. long. Inflorescence a pedunculate, dense, few-flowered capitate cyme, occasionally reduced to a single flower; peduncle terete, 3–13 cm. long; bracteoles lanceolate or elliptic-lanceolate or narrowly ovate-lanceolate, usually narrowed at the base, 2–3.5 cm. long, hairy; pedicels very short or absent. Sepals unequal, 18–3·5 cm. long, outer ones ovate-lanceolate to oblong, inner ones narrower, all usually acute to long acuminate, hairy. Corolla funnel-shaped, magenta, 4–7·5 cm. long; midpetaline areas silky-pilose. Capsule and seeds unknown.

  Botswana. N: Ngamiland, pan nr. Francistown-Maun Rd., 900 m., fl. & fr. 11.iii.1961, *Richards* 14631 (K). SW: 4 km. NE. of Tshobokwane, fl. & fr. 9.xi.1978, *Skarpe* 298 (K). **Zimbabwe.** C: Harare, Mount Pleasant, 4.xii.1966, *Grosvenor* 299 (K; LISC). E: Mutare, Hondi Valley, Souldrop Farm, fl. & fr. 11.ii.1949, *Chase* 1158A (BM; K).
  Also in Namibia, S. Africa (Transvaal). Open savanna, grassland, sandy soil; 900–1435 m.

27. **Ipomoea heterotricha** F. Didr. in Kjoeb. Vidensk. Meddel. 1854): 220 (1854.—Verdc. in F.T.E.A., Convolvulaceae: 107 (1963).—Heine in F.W.T.A. ed 2, **2**: 347 (1963). Type from Guinea.
    Ipomoea amoena Choisy in DC., Prodr. **9**: 365 (1845) *nom. illegit.* non Blume (1825). —Hall. f. in Bot. Jahrb. **18**: 133 (1893).—Hiern, Cat. Afr. Pl. Welw. **1**: 734 (1898). —Baker & Rendle in F.T.A. **4**, 2: 154 (1905).—Hutch. & Dalz., F.W.T.A. **2**: 218 (1931). Type from "Africa and Galam et Cape Coast".
    Ipomoea amoenula Dandy in F.W. Andr., Fl. Pl. Anglo-Egypt. Sudan **3**: 112 (1956). Type as for *Ipomoea amoena* Choisy.

  Annual herb. Stems prostrate to twining up to 1·5 m. long, densely covered with fine soft spreading golden hairs, associated in the younger parts as also in the petiole and peduncle with short appressed white hairs. Leaf lamina ovate, up to 9·5 × 5·5 cm., often more or less cordate at the base, acute at the apex, green but with long appressed hairs above, silvery-velvety with appressed whitish silky hairs beneath; veins and margin also pilose; petiole 1–6 cm. long, densely pilose like the young shoots and peduncles. Flowers in dense cymose heads, up to 4 cm. wide, densely hairy; peduncle 1–7 cm. long; outer bracts foliaceous, bluntly linear-lanceolate, up to 3 cm. long; inner bracts narrow up to 1·8 cm. long, all with an indumentum similar to that on the leaves. Sepals subequal, accrescent, linear-lanceolate, more or less 6 mm. long, densely clothed with long stiffish yellowish hairs. Corolla funnel-shaped, purple or white, up to 2·8 cm. long; midpetaline areas with long stiffish hairs on the upper part of the limb. Capsule ovoid, glabrous. Seeds black, puberulous.

  Zambia. N: Mbala Distr., Kumbula Isl., Lake Tanganyika, 853 m., fl. & fr. 4.v.1952, *Richards* 1597 (K). **Malawi.** N: Mzimba Distr., c. 4·8 km. S. of Mzuzu at Katoto, 1372 m., fl. &

fr. 13.i.1973, *Pawek* 6358 (K; SRGH). S: Machinga, Liwonde National Park, 533 m., fl. & fr.
17.iv.1980, *Blackmore, Brummitt & Banda* 1234 (K).
Also in Sudan, Uganda, Tanzania, Senegal, Cape Verde Isl., Zaire and Angola.
Woodland, grassland, sandy and rocky soil; 850–1380 m.

28. **Ipomoea wightii** (Wall.) Choisy in Men. Soc. Phys. Genève **6**: 470 (1834); in DC. Prodr. **9**:
    364 (1845).—Wight, Icon. t. 1364 (1848).—Peters, Reise Mossamb., Bot. **1**: 239 (1861).
    —C.B. Clarke in Hook. f., Fl. Br. Ind. **4**: 203 (1883).—Hall. f. in Bot. Jahrb. **18**: 133
    (1893).—Dammer in Eng., Pflanzenw. Ost-Afr. **C**: 332 (1895).—Hall. f. in Engl., Bot.
    Jahrb.: 32 (1899).—Baker & Rendle in F.T.A. **4**, 2: 157 (1905).—Schinz, Pl. Menyhart.:
    69 (1905).—Rendle in Journ. Linn. Soc. Bot. **40**: 150 (1911).—Eyles in Trans. Roy. Soc.
    S. Afr. **5**: 456 (1916).—Gomes e Sousa, in Bol. Soc. Est Mocambique **32**: 86 (1936).
    —Suessenguth & Merxm. in Trans Rhod. Sci, Assoc. **43**: 42 (1951).—Brenan in Mem. N.
    Y. Bot. Gard. **9**: 7 (1954).—Meeuse in Bothalia **6**: 737 (1958).—Verdc. in Kew Bull. **13**:
    207 (1958); in F.T.E.A., Convolvulaceae: 110, fig. 22, 2 (1963).—Amico & Bavazzano, in
    Erbario Trop. Firenze No. **6**: 278 (1968).—Binns, H.C.L.M.: 40 (1968).—Ross, Fl.
    Natal: 296 (1972).—Jacobsen in Kirkia **9**: 171 (1973).—Compton, Fl. Swaziland: 478
    (1976). Type from India.

Var. **wightii**
    Convolvulus wightii Wall., Pl. As. Rar. **2**: 55, t. 171 (1831). Type as above.
    Ipomoea arachnoidea Boj., Hort. Maurit.: 228 (1837) *nom. nud.* and ex Choisy in DC.,
    Prodr. **9**: 364 (1845) pro parte.—Hall. f. in Engl., Bot. Jahrb. **18**: 133 (1893). Type from
    Zanzibar.

A very variable perennial herbaceous climber. Stems twining or prostrate when no
support is available, up to 3.5 m. long, covered with more or less spreading yellowish
hairs. Leaf lamina ovate-cordate up to 12 × 12·5 cm., entire or shallowly to deeply
3-lobed, with lobes acute and margins entire or crenate, green, more or less densley
strigose above, more or less densely covered with a floccose-arachnoid tomentum
beneath and veins clearly marked out; petiole up to 8 cm. long, patently or retrorsely
pilose. Flowers in few to many-flowered, dense to somewhat lax heads, very rarely
solitary; peduncle 7–21 cm. long, hairy like the stems and petioles; bracteoles
lanceolate, acuminate, hirsute, with short glandular hairs on the sides and margins,
10–12 mm. long; cyme-branches very short, pedicels absent or nearly so. Sepals linear
to ovate-lanceolate, 6–15 mm. long obtuse or acuminate, with 3 kinds of indumentum,
white-cottony tomentum, appressed or spreading yellow bristles and sessile or stalked
marginal glands, (one or more of these kinds may be absent). Corolla funnel-shaped,
magenta or mauve 2–3·5 cm. long, glabrous or more or less yellow-bristly above.
Capsule globose, slightly to densely bristly above or with a faint white tomentum. Seeds
ovoid, black, very shortly pubescent or glabrous.

Zambia. E: Chipata, fl. & fr. 13.x.1967, *Mutimushi* 2311 (K). **Zimbabwe**. N: Mazoe, 1220
m., fl. & fr. 10.vi. 1946, *Wild* 1119 (K; SRGH). W: Matobo, Farm Besna Kobila, 1463 m., fl. &
fr. iii.1957, *Miller* 42434 (K; SRGH). C: Marondera, Delta Farm 1372 m., fl. & fr. 22.iv.1942,
Dehn 197 (K; M; SRGH). E: Mutare, Imbesa Valley, 1250 m., fl. & fr. 8.iii.1954, *Chase* 5206
(BM; K; SRGH). S: Masvingo Distr., Great Zimbabwe National Park, fl. & fr. 29.iii.1973,
Chiparawasha 641 (SRGH). **Malawi**. N: Chitipa, Misuku Hills, Mugesse, 1800 m., fl. & fr.
14.vii.1970, *Brummitt* 12104 (K; MAL; LISC; SRGH). C: Lilongwe, Dzalanyama Forest
Reserve, Chiunjiza Rd., 1230 m., fl. & fr. 23.iii.1970, *Brummitt* 9258 (K; MAL; SRGH). S:
Blantyre Distr., Ndirande Mt., 1400 m., fl. & fr. 31.v.1970, *Brummitt* 11185a (K; LISC; MAL;
PRE; SRGH). **Mozambique**. N: Maçondes, between Mueda and Chomba, fl. & fr. 25.ix.1948,
Barbosa 2241 (LISC; LMA; WAG). Z: Guruè Mt., 1200 m., fl. & fr. 29.vi.1943, *Torre* 5619 (C;
LISC). T: Angónia, nr. Ulongue, fl. & fr. 10.vii.1979, *Macuácua & Stefanesco* 912 (LMA). MS:
Gorongosa Mt., 1370 m., fl. & fr. 10.vii.1969, *Leach & Cannell* 14295 (LISC; SRGH). M:
Namaacha, Goba, fl. & fr. 18.viii.1967, *Gomes e Sousa & Balsinhas* 4931 (K; LMA).
    Also in Uganda, Kenya, Tanzania, Zanzibar S. Africa, Madagascar and tropical Asia. Forest,
bushland, grassland, riversides and sandy soil; 450–1900 m.
    The var. *kilimandschari* (Dammer) Verdc. occurs in Kenya and Tanzania. The var. *obtu-
sisepala* Verdc. occurs in Tanzania.

29. **Ipomoea pes-tigridis** L., Sp. Pl.: 162 (1753).—Choisy in DC., Prodr. **9**: 363 (1845).
    —Peters, Reise Mossamb., Bot. **1**: 239 (1861).—Peter in Engl. & Prantl, Pflanzenfam. ed.
    1, **4**, 3a: 32 (1891).—in Engl., Bot. Jahrb. **18**: 134 (1893). et **28**: 34 (1899).—Dammer in
    Engl., Pflanzenw. Ost.-Afr. **C**: 332 (1895).—Hall. f. in Bull. Herb. Boiss. **6**: 539
    (1898).—Hiern, Cat. Afr. Pl. Welw. **1**: 735 (1898).—Baker & Rendle in F.T.A. **4**, 2: 158
    (1905).—Schinz, Pl. Menyarth.: 69 (1905).—Eyles in Trans. Roy. Soc. S. Afr. **5**: 455
    (1916).—Gomes e Sousa, Bol. Soc. Est. Moçamb. **32**: 86 (1936).—van Ooststr. in Blumea

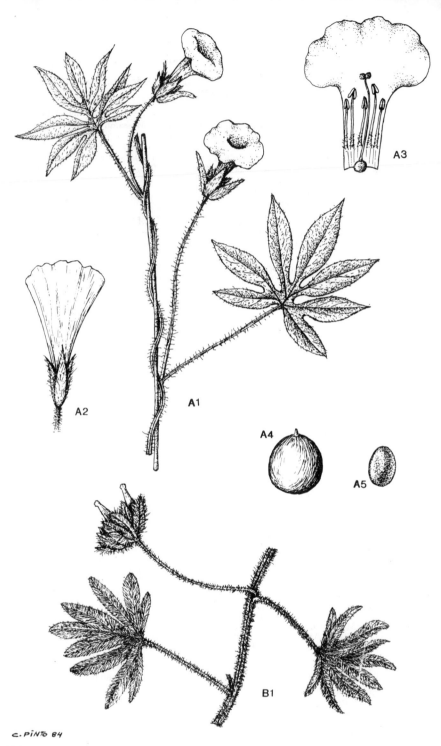

C. PINTO 84

Tab. 21. IPOMOEA PES-TIGRIDIS var. PES-TIGRIDIS. 1, habit (×½); A2, flower (×1); A3, corolla opened to show stamens and pistil (×1), A1–3 from *Goodier* 83; A4, fruit (×1½); A5, seed (×1½), A4–5 from *Fanshawe* 2229; B1, portion of branch of I. PES-TIGRIDIS var. STRIGOSA from *Drummond & Williamson* 9627.

3: 504 (1940); in Fl. Males., Ser. 1, 4: 467, fig. 40 (1953).—Wild, Guide Fl. Victoria Falls: 156 (1953).— Brenan in Mem. N.Y. Bot. Gard. 9: 7 (1954).—Dandy in F. W. Andr., Fl. Pl. Anglo-Egypt. Sudan 3: 113 (1956).—Macnae & Kalk, Nat. Hist. Inhaca Is., Moçamb.: 152 (1958).—Meeuse in Bothalia 6: 744 (1958).—Verdc. in F.T.E.A., Convolvulaceae: 108 (1963).—Heine in F.W.T.A. ed. 2, 2: 347 (1963).—Roessler in Merxm. Prodr. Fl. Sw. Afr. 116: 16 (1967).—Binns, H.C.L.M.: 40 (1968).—Amico & Bavazzano in Erb. Trop. Firenze 49: 542 (1978).—Munday & Forbes in Journ. S. Afr. Bot. 45: 9 (1979). TAB. 21. Type from Sri Lanka.

Annual or sometimes perennial herb. Stems slender or more or less stout, trailing, up to 3 m. long, occasionally prostrate covered with long spreading yellow bristly hairs as are petioles and peduncles, up to 2 m. long. Leaf lamina circular to somewhat reniform in outline, palmately-pedately 7–9-lobed, 4–11 × 4–15 cm., in outline, broadly cordate at the base; lobes narrowly oval to obovate, narrowed above and below, acute or subacute, minutely mucronate at the apex, narrowed and confluent at the base, 1·5–7·5 × 1–2·5 cm., lateral ones gradually smaller, lowermost somewhat oblique to falcate, often obtuse, all thinly hairy to strigose on both surfaces; sinises between the lobes rounded; petiole up to 14 cm. long. Flowers in few-flowered bracteate heads; peduncle up to 15 cm. long; bracts foliaceous, oblong to oblong-lanceolate, hairy outside; outer ones up to 3 × 0·8 cm., inner ones smaller. Sepals lanceolate, 7–12 mm. long, hairy. Corolla funnel-shaped, white, pink or purple with darker throat, 3–5 cm. long; midpetaline areas with a few hairs. Capsule ovoid, glabrous. Seeds brown, pubescent.

Stems slender; plant setose with long hairs - - - - - - - var. *pes-tigridis*
Stems stout; plant very densely setose with long and short hairs, all parts strigose
var. *strigosa*

Var. **pes-tigridis**. Verdc. in F.T.E.A., Convolvulaceae: 108 (1963).
Ipomoea pes-tigridis L. var. *africana* Hall. f. in Bull. Herb. Boiss. 6: 539 (1898). Type from Zanzibar.

Botswana. N: Mboma Camp, fl. & fr. 14.ii.1974, *Smith* 815 (K; SRGH). **Zambia**. B: Sesheke, 915 m., fl. & fr. i.1924, *Borle* 6 (K; PRE). N: Cassava Sands, Lake Tanganyika, 1050 m., fl. & fr. 15.iv. 1957, *Richards* 9241 (K). W: Kitwe, fl. & fr. 25.iii.1955, *Fanshawe* 2229 (K; SRGH). E: Msoro, c. 80 km. W. of Chipata, Luangwa Valley, 732 m., fl. & fr. 11.vi.1954, *Robinson* 857 (K). S: Nr. Zeze, Sinazongwe, 600 m., fl. & fr. 29.xii.1958, *Robson & Angus* 988 (K; LISC; SRGH). **Zimbabwe**. W: Hwange, Kazungula, 915 m., fl. & fr. iv.1955, *Davies* 1129 (SRGH). E: Chimanimani, Haroni Lusitu Junction, fl. & fr. 12.i.1969, *Mavi* 906 (K; LISC). S: Mwenezi, Fishans, Lundi R., fl. & fr. 26.iv.1962, *Drummond* 7739 (K; SRGH). **Malawi**. N: Likoma Isl., Lake Malawi, fl. & fr. 30.iii.1960, *Eyles* 12 (LISC; SRGH). C: Nkhota Kota, 488 m., fl. & fr. 14.ii.1944, *Benson* 312 (PRE). S: W. of Mangochi, fl. & fr. 14.iii.1955, *Exell, Mendonça & Wild* 853 (BM; LISC; SRGH). **Mozambique**. N: Erati, Namapa, Experimental Station C.I.C.A., fl. & fr. 15.iii.1960, *Lemos & Macuácua* 32 (K; LISC; LMA). Z: Mocuba, Namagoa, 61 M., fl. & fr. iii.1943, *Faulkner* 159 (K; PRE; SRGH). T: Cahobra Bassa, Estima-Songo Rd., nr. Marueira, fl. & fr. 5.ii.1972, *Macedo* 4778 (LISC; LMA; LMU) MS: Cheringoma, Inhamitanga, fl. & fr. 6.iv.1945, *Simão* 344 (LISC; LMA). GI: Gaza, Inhamissa, fl. & fr. 24.ii.1955, *Soares Rocha* 54 (LMA). M: Maputo, Chinhambanine, fl. & fr. 21.vi.1965, *Marques* 561 (COI; LMA).
Widely distributed in west, east and S. Africa, also in the Mascarene Isl., tropical Asia and Malaysia. Forest, savanna, bushland, grassland, riverside, cultivated ground, sandy soil, semi-ruderal; 0–1050 m.

Var. **strigosa**. (Hall. f.) Rendle in F.T.A. 4, 2: 159 (1905). Type from Angola.
Ipomoea pes-trigridis L. subvar. *strigosa* Hall. f. in Engl., Bot. Jahrb. 28: 34 (1809).

Zambia. W: Mufulira, fl. & fr. 11.iii.1974, *Chisumpa* 143 (K). C: Serenje, 8 km. SE. of Kanona, on Rd., to Kundalila Falls, fl. & fr. 16.ii.1970, *Drummond & Williamson* 9627 (SRGH). Also in Tanzania, Angola and Zaire. Woodland, grassland, damp sandy ground; c. 1000 m.

30. **Ipomoea ficifolia** Lindl. in Bot. Reg. 26, Misc. Not.: 90 (1840) et 27, t. 13 (1841).—Hall. f. in Engl., Bot. Jarhb. 18: 135 (1893); 28: 35 (1899).—Baker & Wright in Dyer, F.C. 4: 64 (1904).—Macnae & Kalk, Nat. Hist. Inhaca Isl., Moçamb.: 152 (1958).—Meeuse in Bothalia 6: 738 (1958).—Verdc. in F.T.E.A., Convolvulaceae: 111 (1963).—Ross, Fl. Natal: 296 (1972).—Compton, Fl. Swaziland: 477 (1976).—Munday & Forbes in Journ. S. Afr. Bot. 45: 9 (1979). Type a plant cultivated in the Victoria Nursery, Bath (England) Nov., 1840 (CGE, holotype).
Convolvulus trilobus Thunb., Prodr. Fl. Cap.: 35 (1794), nom. illegit. non Desr. (1792). Type from S. Africa.

Ipomoea engleriana Dammer in Engl., Pflanzenw. Ost-Afr. **C**: 333 (1895). Type from Tanzania.

Ipomoea ficifolia Lindl. var. *laxiflora* Hall. f. in Sitz. Akad. Wiss. Wien. **107**: 48 (1898).—Baker & Rendle in F.T.A. **4**, 2: 161 (1905).—Brenan, T.T.C.L.: 170 (1949). Type as for *Ipomoea engleriana.*

Ipomoea ficifolia Lindl. subvar. *auriculata* Hall. f. in Ann. Ist. Bot. Roma **7**: 230 (1898). Type from Somalia.

Perennial from a tuberous root. Stems prostrate or twining, slender up to 2·5 m. long, densely bristly pubescent. Leaf lamina ovate-cordate to subcircular-reniform in outline, 2–6·5 × 1·5–5 cm., entire or 3-lobed, acute or acuminate at the apex, thin, green, appressed pilose above, with dense to very sparse and almost non exsistent traces of white cottony tomentum beneath; veins yellow-bristly beneath; mid-lobe elliptic and mucronulate, basal lobes rounded, all constricted at the base and more or less crenate; petiole 1·5–5 cm. long, slender, hairy as in the stems. Flowers few- to many, in lax, rarely rather dense, cymes, very rarely solitary; peduncle up to 7 cm. long, slender, pilose like the stems and petioles; bracteoles linear-lanceolate, 5–11 mm. long, acuminate-aristate, pilose or hirsute as are the sepals; cyme branches usually short; pedicels 0–10 mm. long. Sepals linear-lanceolate to lanceolate, very acuminate, 6–18 mm. long, pilose with yellow-bristly hairs; margins glandular. Corolla broadly funnel-shaped, pink or mauve, 4.5–5 cm. long, glabrous or with a few pilose hairs on the well-defined mid-petaline areas; the limb shallowly 5-lobes. Capsule ovoid, glabrous or pubescent. Seeds compressed-globose, yellow-pubescent, also with 2 tufts of long white hairs attached to edges near the apex.

Mozambique. GI: Chongoene Beach, fl. & fr. 19.v.1966, *Balsinhas* 1083 (COI; LMA). M: Maputo, Inhaca Isl., fl. & fr. 31.viii.1959, *Watmough* 366 (K; SRGH).

Also in Somalia, Kenya, Tanzania and S. Africa. Open woodland, littoral scrub, grassland and coast sandy soil; 0–200 m.

31. **Ipomoea dichroa** Hochst. ex Choisy in DC., Prodr. **9**: 364 (1845).—Verdc. in Kew Bull. **33**: 165 (1978). Type from Senegal.

Convolvulus bicolor Desr. in Lam., Encycl. **3**: 564 (1791) non vani (1794), non *Ipomoea bicolor* Lam. Type as for *Ipomoea dichroa.*

Convolvulus dichrous Roem. & Schultes, Syst. Veg. **4**: 263 (1891) *nom. illegit.* Type as above.

Convolvulus pilosus Roxb., Fl. India, ed. Carey & Wall. **2**: 55 (1824) *nom illegit.* non Rottler (1803).

Ipomoea pilosa (Roxb.) Sweet, Hort. Brit.: 289 (1827) *nom. illegit.* non Houttuyn (1777).—Hiern, Cat. Afr. Pl. Welw. **1**: 735 (1898).—Baker & Rendle in F.T.A. **4**, 2: 161 (1905).—Schinz Pl. Menyhart.: 69 (1905).—N.E. Br. in Kew Bull. **1909**: 122 (1909). —*Gomes e Sousa* in Bol. Soc. Est. Moçambique **32**: 86 (1936).

Ipomoea arachnosperma Welw., Phyto-Geogr. Apont.: 588 (1859).—Meeuse in Fl. Pl. Afr. **31**: t. 1203 (1956); in Bothalia **6**: 736 (1958).—Verdc. in F.T.E.A., Convolvulaceae: 112 (1963).—Roessler in Merxm., Prodr. Fl. SW. Afr. **116**: 13 (1967).—Binns, H.C.L.M: 40 (1968).—Jacobsen in Kirkia **9**: 171 (1973). Type from Angola.

Annual resembling *Ipomoea ficifolia* Lindl. but with smaller flowers. Stems several from the base, twining or occasionally prostrate, up to 2·5 m. long, densely covered with white or yellow bulbous-based patent bristly hairs. Leaf lamina broadly cordate-ovate in outline, up to 18·5–17·5 cm., deeply 3-lobed, rarely entire, acute or acuminate at the apex, thin, green, bristly-pubescent above, covered with dense white-cottony tomentum beneath; petiole 4–10 cm. long. Flowers several in lax cymes; peduncle 1·5–7 cm. long; secondary and tertiary branches 1·5–3 cm. long; bracts ovate-lanceolate with a broad base, long-acuminate to aristate, up to 11 mm. long; pedicels, somewhat flattened, up to 1 cm. long. Sepals subequal, lanceolate, long acuminate or aristate, 1·2–2·5 cm. long, bristly or setose and glandular, accrescent in fruit. Corolla funnel-shaped, pink or mauve with darker mauve centre, 1·4–2·0 cm. long, pubescent on midpetaline areas near the tips of the corolla lobes. Capsule ovoid, pubescent above. Seeds ovoid, black, velvety-white-pubescent.

Botswana. N: Dindings Isl., fl. & fr. 28.iii.1975, *Smith* 1312 (K; SRGH). **Zambia.** B: Sesheke Distr., Kazu Forest, 1180 m., fl. & fr. 9.iv.1955, *Exell, Mendonça & Wild* 1450 (BM; LISC; SRGH). W: Ndola, fl. & fr. 3.iv.1954, *Fanshawe* 1044 (K; SRGH). C: Mt. Makulu Research Station, 16 km. S. of Lusaka, fl. & fr. 31.iii.1960, *Smartt* 4 (K; LISC; SRGH). S:

Lusitu, fl. & fr. 19.v.1960, *Fanshawe* 5680 (K; SRGH). **Zimbabwe. N**. Mazoe, 1280 m., fl. & fr. iv.1907, Eyles 536 (BM; K; SRGH). W: Hwange, Makwa Pan, 14 km. SE. of Main Camp, Hwange National Park, fl. & fr. 17.iv.1972, *Grosvenor* 701 (K; SRGH). C: Harare, 1463 m., fl. & fr. 27.iii.1929, *Eyles* 4863 (K; SRGH). S: Beitbridge, Shashi R., c. 3 km. upstream from Tuli Police Camp, fl. & fr. 14.v.1959, *Drummond* 6144 (K; SRGH). **Malawi.** N: from Kondowe to Karonga, 610–1830 m., fl. & fr. vii.1896, *Whyte* s.n. (K). S: Shire R., fl. & fr. 1863, *Kirk* s.n. (K). **Mozambique. T**: Tete, from Estima to Chinhanda, 1·5 km. from Estima, fl. & fr. 10.iv.1972, *Macedo* 5176 (K; LISC; LMA; LMU; SRGH). MS: Chemba, Chiou, Exper. Stat. of C.I.C.A., fl. & fr. 12.iv.1960, *Lemos & Macuácua* 82 (BM; COI; K; LISC; LMA; SRGH). GI: Limpopo, Chicualacuala, fl. & fr. 26.iii.1974, *Balsinhas & Santos* 2655 (LMA).

Throughout most of tropical Africa and India; extends into Namibia and S. Africa (Transvaal). Riverine forest, secondary savanna, grassland, cultivated ground, roadside and sandy soil; 0–1830 m.

32. **Ipomoea pharbitiformis** Baker in Kew Bull. **1895**: 291 (1895); in F.T.A. **4**, 2: 162 (1905).—Verdc. in Kew Bull. **13**: 208 (1958) & 14: 341 (1960).—White, F.F.N.R.: 362 (1962).—Verdc. in F.T.E.A., Convolvulaceae: 112 (1963). Type: Zambia, Lake Mweru, *Carson* 41 (1894 coll.) (K, holotype).

Ipomoea verdickii De Wild., Études Fl. Katanga: 113, t. 3, fig. 9–16 (1903). Type from Zaire.

A robust rather woody climber, up to 15 m. Stems slender, climbing, pubescent, roughly hirsute in longitudinal lines or almost glabrous; older ones woody, pale brown and ridged. Leaf lamina ovate 3·5–18 × 2·5–17 cm. acute or acuminate at the apex, cordate at the base, green and thinly pilose or glabrous above, glabrescent to densely hairy beneath with matted more or less appressed hairs but never with arachnoid tomentum; petiole 1–13 cm. long, with a double line of pale brownish soft spreading hairs. Flowers in rather lax to dense, few- to many-flowered cymes; peduncle 1·5–10 cm. long; pedicels up to 2·5 cm. long; bracts lanceolate to ovate-lanceolate, acute up to 13 mm. long appressed pilose and densely glandular. Sepals similar to the bracts in shape and indumentum, 8–12 mm. long. Corolla funnel-shaped, mauve with a deeper mauve centre, 3–4 cm. long; midpetaline areas sparsely pubescent. Capsule globose, pilose. Seeds subglobose, blackish-brown, glabrous.

Zambia. N: Mbala Distr., Kambole Escarpment, 1500 m., 23.viii.1956, *Richards* 5926 (K). W: Mwinilunga Distr., Lisombo R., 8.vi.1963, *Edwards* 689 (K; LISC; SRGH).

Also in Tanzania and Zaire. Swamp forest, banks of river and damp sandy soil; 1500 m.

33. **Ipomoea magnusiana** Schinz in Verh. Bot. Brandenb. **30**: 272 (1888).—Hall. f. in Engl., Bot. Jahrb. **18**: 135 (1893).—Baker & Wright in Dyer, F.C. **4**: 65 (1904).—Baker & Rendle in F.T.A. **4**, 2: 162 (1905).—N.E. Br. in Kew Bull. **1909**: 123 (1909).—Eyles in Trans. Roy. Soc. S. Afr. **5**: 455 (1916).—Dinter in Fedde, Repert **18**: 431 (1922).—Meeuse in Bothalia **6**: 742 (1958).—Roessler in Merxm., Prodr. Fl. SW. Afr. **116**: 15 (1967).—Ross, Fl. Natal: 296 (1972).—Compton, Fl. Swaziland: 478 (1976).—Munday & Forbes in Journ. S. Afr. Bot. **45**: 9 (1979). Type from Namibia.

Perennial, forming several to many annual stems from a woody taproot. Stems twining, climbing, prostrate or suberect young ones, up to several metres, slender or stouter, pilose. Leaf lamina palmately 3- or 5-lobed nearly to the base, 2–11 × 2–14 cm., green or yellow-green, with thin appressed pilose hairs above, densely covered with a white cobwebby tomentum beneath except on the main nerves and main veins which are covered with yellowish or brownish stiff, appressed to patent hairs and thus clearly marked out; lobes varying from obovate to narrowly elliptic, lanceolate, oblanceolate, linear-lanceolate or ovate-lanceolate except the basal ones in 5-lobed leaves which are shorter and relatively broader, apex or central and first pair of lobes subacute to acuminate, often cuspidate; margins entire or subentire, more or less distinctly ciliate; petioles more or less as long as the leaves, slender or stouter, pilose. Inflorescence a dense few-flowered pedunculate head, rarely (by reduction) flowers solitary; peduncle up to 12 cm. long, slender or stouter, pilose as in stems and petioles; bracts linear or lanceolate up to 15 mm., hairy; pedicels very short. Sepals somewhat unequal, lanceolate, acute, 6–15 mm. long, accrescent in fruit up to 20 mm., hairy outside. Corolla funnel-shaped, magenta-purple, mauve, cream, or white, with darker magenta centre, 12–32 mm. long; midpetaline areas pilose. Capsule subglobose, glabrous. Seeds pubescent, with the hairs often arranged in small tufts and sometimes with long white hairs on the angles in upper half.

Stems twining, prostrate or suberect (young ones) up to about 2 m. long, slender; leaf lamina 2–6 × 2–6 cm.; petioles and peduncles slender; corolla 12–20 mm. long - var. *magnusiana*
Stems usually climbing, up to several metres, generally stouter; leaf-lamina 4–11 × 4–14 cm.; petioles and peduncles stouter; corolla 20–32 mm. long - - - - - var. *eenii*

Var. **magnusiana** Meeuse in Bothalia **6**: 742 (1958).
　　Ipomoea lugardii N.E. Br. var. *parviflora* Rendle in F.T.A. **4**, 2: 163 (1905). Type: Zimbabwe, Matabeleland, ii.1866, *Elliott* (K, holotype).
　　Ipomoea otjikangensis Pilger & Dinter in Engl., Bot. Jahrb. **41**: 296 (1908).—Dinter in Feddes Repert. **18**: 431 (1922). Type from Namibia.

Botswana. N: Ngamiland, Makarikari, 930 m., fl. & fr. 9.iii.1961, *Richards* 14656 (K; SRGH). SW: Ghanzi, Eaton's Farm, fl. & fr. 9.v.1969, *Brown* 6037 (K; SRGH). SE: Mochundi, 946, fl. & fr. i.1914, *Rogers* 6521 (K; PRE). **Zimbabwe**. W: Bulawayo, 1361 m., fl. & fr. 18.ii.1912, *Rogers* 5748 (K). E: Chipinge, nr. Sabi at Dott's Drift, fl. & fr. 17.ii.1960, *Goodier* 924 (K; SRGH). S: Ndanga, c. 19 km. S. of Mukwasini R., 487 m., fl. & fr. 28.i.1957, *Phipps* 187 (COI; SRGH). **Mozambique**. N: Montepuez, Nairó, fl. & fr. 27.viii.1972, *Mafumo* 18 (LISC). GI: Caniçado, Massingir, between Lagoa Nova and Mongonso, 15 km. from Lagoa Nova, fl. & fr. 17.iii.1972, *Myre, Lousã & Rosa* 5714 (LMA). M: Maputo, Inhaca Isl., fl. & fr. 22.ix.1957, *Mogg* 27458 (K; SRGH).
Also in Namibia and S. Africa. Forest, savanna, grassland, roadside and sandy soil; 120–1360 m.

Var. **eenii** (Rendle) Meeuse in Dyer, Fl. Pl. Afr. **31**: pl. 1201 (1956); Bothalia **6**: 743 (1958). Type from Namibia.
　　Ipomoea eenii Rendle in Journ. Bot. **39**: 21 (1901); in F.T.A. **4**: 163 (1905), excl. var. *parviflora*.
　　Ipomoea lugardii N.E. Br. in F.T.A. **4**: 163 (1905); in Kew Bull. **1909**: 124 (1909). Type: Botswana, Ngamiland, Kgwebe Hills, *Lugard* 211 (K, holotype).

Botswana. N: Ngamiland, between Lake Ngami and Kgwebe Hills, fl. & fr. 17.ii.1966, *Drummond* 8760 (SRGH). SE: Kgatla, 10 km. NW. of Sikwane, 1067 m., fl. & fr. iv.1955, *Reyneke* 206 (K; PRE). **Zimbabwe**. S: Brichenough Bridge, Sabi R., fl. & fr. i.1938, *Obermeyer* 2496 (PRE). **Mozambique**. M: nr. Maputo, fl. & fr. 4.i.1941, *Torre* 2459 (C; LISC; LMA; WAG).
Also in Namibia and S. Africa. Scrubland-woodland, savanna and sandy soil; 950–1067 m.

34. **Ipomoea indica** (Burm.) Merr., Interpr. Rumph. Herb. Amboin: 455 (1917).—Fosberg, Micronesica **2**: 151 (1967); in Bot. Notiser **129**: 37 (1976). Type from Indonesia.
　　Convolvulus indicus Burm. in Rumph. Herb. Amboin., Index Universalis **7**: 6 (1755).
　　Ipomoea congesta R. Br., Prodr.: 485 (1810).—van Ooststr. in Fl. Males., Ser. 1, **4**: 465, fig. 39a (1953).—Verdc. in Taxon **6**: 231 (1957).—Meeuse in Bothalia **6**: 735 (1958). Type from Australia.

Var. **acuminata** (Vahl) Fosberg in Bot. Notiser **129**: 38 (1976). Type from the West Indies.
　　Convolvulus acuminata Vahl, Symb. Bot. **3**: 26 (1794).
　　Ipomoea acuminata (Vahl) Roem. & Schultes, Syst. **4**: 228 (1819).—Verdc. in F.T.E.A., Convolvulaceae: 113 (1963).

Ornamental herbaceous climber, at least sometimes perennial. Stems twining or prostrate up to several metres long, rooting at the nodes, pilose to glabrescent. Leaf lamina ovate in outline, 5–12 × 4·5–15 cm., entire or prominently 3-lobed, cordate at the base, lobes acuminate, pilose to glabrescent; petiole 2–10 cm. long. Flowers few to several in axillary cymes; peduncle 0·5–15 cm. long; pedicels 2–10 mm. long; bracts linear to lanceolate or ovate-lanceolate. Sepals lanceolate 1·5–2·3 cm. long, caudate-acuminate, herbaceous, glabrescent. Corolla funnel-shaped, blue or mauve-purple, often red-tinged, 5–8 cm. long, glabrous; tube whitish at the base; limb flaring. Capsule globose, glabrous. Seeds black, glabrous.

Zambia. C: Lusaka, Forest Nursery, 1961, *Fanshawe* s.n. (SRGH). **Zimbabwe**. N: Mazoe, 1220 m., fl. 16.iii.1971, *Searle* 196 (K; LISC; SRGH). E: Mutare, 1100 m., fl. & fr. 9.ix.1959, *Chase* 7164 (SRGH). **Malawi**. N: Mzimba, Marymount, Mzuzu, 1370 m., fl. & fr. 17.xi.1970, *Pawek* 4008 (K). S: Blantyre Distr., Maone, 2 km. NE. of Limbe, 1200 m., fl. & fr. 13.iii.1970, *Brummitt* 9078 (EA; K; LISC; MAL; PRE; SRGH; UPS).
Pantropical, often cultivated as an ornamental. An introduced plant, naturalised in waste places, granitic sand and roadsides; 1100–1370 m.

35. **Ipomoea nil** (L.) Roth, Cat. Bot. **1**: 36 (1797).—Hall. f. in Engl., Bot. Jahrb. **18**: 136 (1893).—Dandy in F. W. Andr., Fl. Pl. Anglo-Egypt. Sudan **3**: 119 (1956).—Verdc. in Taxon **6**: 231 (1957) et **7**: 84 (1958).—Meeuse in Bothalia **6**: 753 (1958).—Heine in F.W.T.A., ed. 2, **2**: 351 (1963).—Verdc. in F.T.E.A., Convolvulaceae: 113 (1963).

—Ross Fl. Natal: 296 (1972). Type from U.S.A., Virginia & Carolina: illustration of *Convolvulus caeruleus hederaceo folio magis andguloso* in Dill., Hort. Elth, t. 80, fig. 92 (1732) (syn.).
   Convolvulus nil L., Sp. Pl., ed. 2: 219 (1762).
   Ipomoea hederacea sensu auctt. Mult., e.g. Baker & Rendle in F.T.A. 4, 2: 159 (1905) in Part & Verdc. in Taxon 6: 231 (1957) non Jacq.

Ornamental herbaceous annual. Stems twining, bristly-pilose. Leaf lamina ovate to circular in outline, 3-lobed, up to 14 × 13·5 cm., cordate at the base, more or less appressed pilose above and beneath; middle lobe ovate to oblong, acuminate, lateral ones obliquely ovate to broadly falcate, acuminate; petiole up to 8 cm. long, bristly. Flowers solitary or in lax few-flowered cymes; peduncle up to 7 cm. long, hirsute like the stems; bracteoles linear to filiform, 5–10 mm. long; pedicels up to 10 mm. long, retrorsely hairy. Sepals linear-lanceolate, 2·3–2·8 cm. long, long-attenuated at the apex, densely pilose with patent bristles at the base and a few above together with much shorter pubescence. Corolla funnel-shaped, magenta with paler tube, 5·5–7 cm. long, glabrous. Capsule ovoid to globose, glabrous. Seeds black, puberulous.

Zambia. W: Ndola Distr., Sacred Lake, nr. St. Antony's Mission, c. 48 km. SW. of Luanshya, 1200 m., fl. 14.ii.1975, *Hooper & Townsend* 31 (K). S: Mazabuka, Simasunda, Mapanza, 1067 m., fl. & fr. 21.ii.1957, *Robinson* 2138 (K; SRGH). **Zimbabwe**. N: Mazoe, 1280 m., fl. & fr. iv.1907, *Eyles* 537 (BM; K; SRGH). C: Harare, 1463 m., fl. & fr. 10.iii.1927, *Eyles* 4688 (K; SRGH). E: Mutare, Murahwa's Hill, 1280 m., fl. & fr. 13.ii.1963, *Chase* 7955 (K; LISC; SRGH).
   Also in Sudan, Uganda, Tanzania, S. Africa (Natal), North America and Malaysia; introduced into various parts of the tropics. Woodland, grassland and waste ground; 1000–1525 m.

36. **Ipomoea purpurea** (L.) Roth, Bot. Abh.: 27 (1787); Cat. Bot. 1: 36 (1797).—Hall. f. in Engl., Bot. Jahrb. 18: 137 (1893).—van Ooststr. in Fl. Males., Ser. 14: 465 (1953). —Verdc. in Taxon 6: 231 (1957).—Meeuse in Bothalia 6: 734 (1958).—Verdc. in F.T.E.A., Convolvulaceae: 114 (1963).—Ross, Fl. Natal: 296 (1972). TAB. 22. Type from U.S.A.
   Convolvulus purpurea L., Sp. Pl., ed. 2: 219 (1762).

Ornamental herbaceous annual. Stems trailing or twining, glabrous or with short hairs mixed with longer retrorse bristles. Leaf lamina ovate to subcircular in outline, entire or 3-lobed, 4–16 × 3–15 cm., the base cordate with broadly rounded auricles, the apex acuminate, glabrous or pubescent; petiole 6·5–15 cm. long, retrorsely hirsute. Flowers solitary or in few-flowered cymes; peduncle up to 15 cm. long; pedicels shortly hairy or with a few bristles, 8–15 mm. long; Sepals unequal, 10–15 mm. long, accrescent up to 20 mm. long in fruit; outer ones oblong, acute with bristly patent hairs in basal portion; inner ones with narrow scarious margins, linear-oblong to linear, acute with a few bristles near the base. Corolla funnel-shaped, white, pink or magenta 4·5–6 cm. long; tube white below. Capsule globose, glabrous. Seeds black, glabrous or very shortly pubescent.

Zambia. B: Sesheke, fl. & fr. i.1924, *Borle* s.n. (PRE). N: Mbala, 1676 m., fl. & fr. 14.v.1957, *Vesey-FitzGerald* 1234 (SRGH). W: Ndola, 4.ii.1971, *Fanshawe* 11138 (K; SRGH). C: Lusaka, fl. & fr. 19.ii.1957, *Noak* 117 (K; SRGH). **Zimbabwe**. N: Mazoe, fl. & fr. 2.iii.1971, *Searle* 86 (K; SRGH). W: Bulawayo, 1372 m., fl. & fr. iii.1958, *Miller* 5139 (SRGH). C: Harare, junction Drummond Chaplin Rd./West Rd., 1460 m., fl. 18.i.1971, *Biegel* 3453 (K; LISC; SRGH). **Mozambique**. T: Tete, right edge of Zambezi R., fl. 20.x.965, *Neves Rosa* 86 (LISC; LMA).
   Also in Uganda, Kenya, Tanzania and S. Africa; native of S. America, now widely naturalised in the tropics, Malaysia, probably occuring throughout tropical Africa. Woodland, bushland, grassland and waste ground; 1370–1680 m.

37. **Ipomoea parasitica** (Kunth) G. Don, Gen. Syst. 4: 275 (1837).—O'Donall in An. Inst. Biol. Mexico 12: 91, fig. 5 (1941).—Verdc. in Kirkia 6: 121 (1967). Type from Venezuela.
   Convolvulus parasiticus Kunth, Nov. Gen. Sp. Pl. 3: 103 (1818–19).

Ornamental herbaceous climber, well naturalised as an escape from cultivation. Stems trailing for several metres and climbing over low bushes, reddish, muriculate, glabrous or pubescent. Leaf lamina ovate, 5–15 × 5–15 cm., obtuse or acuminate at the apex, cordate, upper surface appressed-pubescent, glabrescent and scarcely pubescent on the nerves beneath; petiole 5–24 cm. long, pubescent or glabrous, sometimes

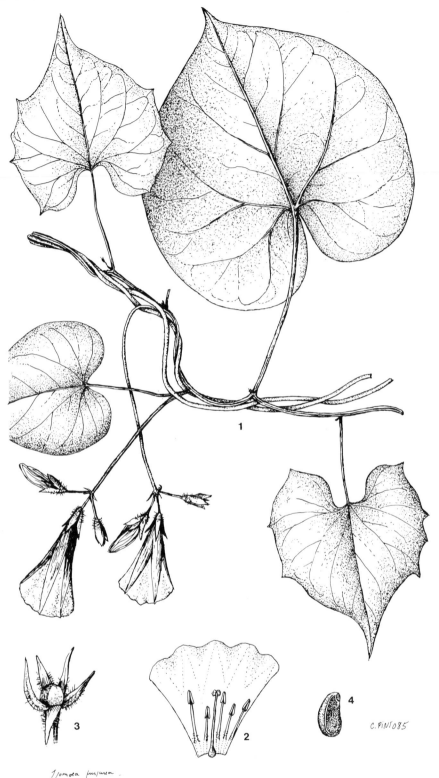

Ipomoea purpurea.

Tab. 22. IPOMOEA PURPUREA. 1, habit ($\times\frac{1}{2}$); 2, corolla opened to show stamens and pistil ($\times\frac{1}{2}$), 1–2 from *Fanshawe* 11138; 3, fruit ($\times$1); 4, seed ($\times$4), 3–4 from *Mutimushi* 1251.

muriculate. Inflorescence several-flowered cymes; peduncle 5–25 cm. long, pubescent; pedicels 0·7–1.5 cm. long, enlarged at the apex; bracts lanceolate, 4–7 mm. long, pubescent, deciduous. Sepals subequal, lanceolate-oblong, 5 mm. long, mucronate, pubescent outside. Corolla funnel-shaped, blue with purplish tinge, 3·5–4·5 cm. long, pubescent at the midpetaline areas. Capsule ovoid to globose, glabrous. Seeds trigonous, brownish, glabrous or very shortly pubescent.

Zambia. S: Machili, fl. 20.iv.1961, *Fanshawe* 6511 (K; SRGH). **Malawi**. S: Zomba, Nandolo Lines, fl. & fr. 7.v.1965, *Salubeni* 332 (K; SRGH). **Mozambique**. N: Nampula, fl. & fr. 21.iv.1937, *Torre* 1410 (COI; LISC).

Cultivated species originally of Venezuela and Central America, escaped from cultivation and well naturalised. Open bushland and pathside banks; 975 m.

38. **Ipomoea batatas** (L.) Lam., Tab. Encycl. Meth. Bot. **1**: 465 (1793).—Peter in Engl. & Prantl, Pflanzenfam. ed. 1, **4**: 3a: 30 (1891).—Hall. f. in Engl., Bot. Jahrb. **18**: 138 (1893).—Dammer in Engl., Pflanzenw. Ost–Afr. **C**: 332 (1895).—Hiern, Cat. Afr. Pl. Welw. **1**: 736 (1898).—Baker & Rendle in F.T.A. **4**, 2: 175 (1905).—Schinz, in Pl. Menyhart: 69 (1905).—Stuhlmann in Deutsch–Ost–Afr. **10**: 246–251 (1909). Gomes e Sousa in Bol. Soc. Est. Moçambique **32**: 86 (1936).—van Ooststr. in Blumea **3**: 512 (1940); in Fl. Males., Ser. 1, **4**: 469 (1953).—Macnae & Kalk, Nat. Hist. Inhaca Isl. Moçambique: 152 (1958). Meeuse in Bothalia **6**: 746 (1958).—Verdc. in F.T.E.A., Convolvulaceae: 114 (1963).—Heine in F.W.T.A. ed. 2, **2**: 350 (1963). TAB. **23**. Type from India.
    Convolvulus batatas L., Sp. Pl.: 154 (1753).
    Convolvulus edulis Thunb., Fl. Japan: 84 (1784). Type from Japan.
    Batatas edulis (Thunb.) Choisy in DC., Prodr. **9**: 338 (1845).—Welw., Ann. Conselho Ultramar Apont. Phyto-Georgr.: 551 (1859).—Peters, Reise Mossamb. Bot. **1**: 238 (1861). Type as for *Convolvulus edulis*.
    Ipomoea batatas (L.) Lam. var. *cannabina* Hall. f. in Engl., Bot. Jahrb. **28**: 37 (1899). Type from Zanzibar.

Perennial plant herbaceous, with underground, fusiform to ellipsoid, yellow or reddish, edible tubers. Stems prostrate, ascending or rarely twining, often rooting at the nodes, containing a milky juice, glabrous or very slightly pubescent. Leaf lamina triangular to broadly ovate in outline, 4–10 × 4–13 cm., entire or palmately shallowly to very deeply 3–7-lobed, truncate or cordate at the base; lobes triangular, lanceolate to linear—oblong, glabrous or slightly pubescent; petiole 3·5–15 cm. long, glabrous or hairy. Inflorescence axillary, cymosely 1 to several-flowered; peduncle stout, 3–15 cm. long, glabrous or hairy, bracteoles minute, narrow, acute, 2–3 mm. long, early deciduous; pedicels 3–12 mm. long. Sepals subequal, subcoriaceous, 7–10 mm. long; outer ones oblong or elliptic-oblong; inner ones elliptic-oblong or ovate oblong, somewhat longer, all glabrous or pilose on the back and fimbriate, acute or subacute, distinctly mucronate. Corolla bell-shaped, pale-mauve, white above, 3–4·5 cm. long. Capsule ovoid. Seeds glabrous.

Zimbabwe. C: Harare, 1524 m., fl. v.1918, *Eyles* 1329 (SRGH). **Malawi**. C: Nkhota Kota, fl. vii.1896, *Webb* s.n. (BM). **Mozambique**. MS: Mossurize, Maringa, fl. vi.1973, *Bond* 20 (SRGH). M: Maputo, Inhaca Isl., 15 m., fl. 11.vii.1957, *Mogg* 27356 (K; LMA; SRGH).

Also in tropical east Africa, Angola and southern Africa, probably originated in South America, distributed throughout all tropical areas and widely cultivated in all suitable areas in the territories. Riverine sands and sandy soils; 0–1525 m. The Sweet Potato is cultivated in areas of moderate rainfall or in wet places.

39. **Ipomoea ochracea** (Lindl.) G. Don, Gen. Syst. **4**: 270 (1837).—Hall. f. in Engl., Bot. Jarhb. **18**: 140 (1893); in Bull. Herb. Boiss. **6**: 540 (1898).—Hiern, Cat. Afr. Pl. Welw. **1**: 737 (1898).—Baker & Rendle in F.T.A. **4**, 2: 166 (1905) pro parte.—N.E. Br. in Kew Bul. **1909**: 122 (1909).—Meeuse in Fl. Pl. Afr. **31**: t. 1221 (1956).—Verdc. in Kew Bull. **13**: 210 (1958); in F.T.E.A., Convolvulaceae: 115 (1963).—Heine in F.W.T.A. ed. 2. **2**: 349 (1963).—Binns, H.C.L.M.: 40 (1968). Type from Ghana.

Perennial herbaceous plant. Stems slender, up to 5 m. long, prostrate or twining, pubescent, villous or subglabrous. Leaf lamina cordate-ovate, up to 10 × 9 cm., acuminate at the apex, entire, membranous, glabrescent or pubescent above, shortly pubescent beneath; petiole up to 9 cm. long, slender. Inflorescence 1–several-flowered lax; peduncle slender, up to 5·5 cm. long; pedicels unequal, up to 4 cm. long, shorter than the peduncle, thickened above; bracteoles minute, ovate, acuminate. Sepals

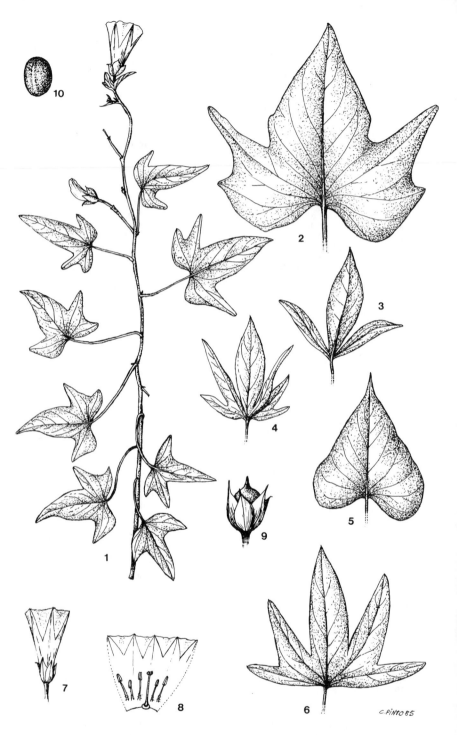

Tab. 23. IPOMOEA BATATAS. 1, habit ($\times\frac{1}{2}$), from *Scott* s.n.; 2, leaf ($\times\frac{1}{2}$), from *Torre* 7677; 3 & 4, leaves ($\times\frac{1}{2}$), from *Mogg* 27356; 5, leaf from *van Rensberg* 2876; 6, leaf ($\times\frac{1}{2}$), from *Astle* 5155; 7, flower ($\times\frac{1}{2}$); 8, corolla opened to show stamens and pistil ($\times\frac{1}{2}$), 7–8 from *van Rensberg* 2876; 9, fruit ($\times1\frac{1}{2}$); 10, seed ($\times2$), 9–10 from *Jansen & Koning* 7395.

ovate, up to 6 mm. long, acute, glabrous or pubescent. Corolla funnel-shaped bright yellow or white, with a dark purple or brown centre, 2·7–5·5 cm. long, glabrous. Capsule globose to conical, glabrous, tipped by the hardened style base. Seeds brown, pubescent.

Var. **ochracea** Verdc. in F.T.E.A., Convolvulaceae: 115 (1963).
  Convolvulus ochraceus Lindl., Bot. Reg. **13**: t. 1060 (1827).
  Ipomoea afra Choisy in DC., Prodr. **9**: 380 (1845). Type from "Guinea".
  Ipomoea kentrocarpa Hochst. ex A. Rich., Tent. Fl. Abyss. **2**: 70 (1851).—Baker & Rendle in F.T.A. **4, 2**: 163 (1905).—Hutch. & Dalz., F.W.T.A. **2**: 216 (1931). Type from Ethiopia.
  Ipomoea opthalmantha Hall. f. in Engl., Bot. Jahrb. **18**: 141 (1893). Type from Tanzania.

  Botswana. N: Ngamiland, Kgwebe Hills, 1036 m., fl. & fr. 18.i.1898, *Lugard* 114 (K).
Zambia. B: Siwelewele, Mashi Rd., 1036 m., fl. & fr. 8.viii.1952, *Codd* 7453 (BM; COI; K; PRE; SRGH). N: Mbala Distr., Rd., to Lunzua Power Station, 900 m., fl. & Fr. 22.v.1962, *Richards* 16492 (K; SRGH). C: Lusaka, Barlaston Park, 1220 m., fl. & fr. 9.x.1969, *Best* 242 (SRGH). S: Mazabuka, Simons Farm, 1006 m., fl. & fr. 8.viii.1931, *Trapnell* 405 (K).
Zimbabwe. N: Urungwe, 20 km. from Makuti, 1080–1200 m., fl. & fr. 18.ii.1981, *Philcox et. al.* 8717 (K). C: Marondera, fl. & fr. xiii.1926, *Rand* 424 (BM). **Malawi.** N: Nkhata Bay Distr., junction N.B. & Kabunduli roads, 762 m., fl. & fr. 3.vi.1973, *Pawek* 6788 (K; SRGH). S: Ntcheu, Dombale-Livulezi Rivers confluence, fl. & fr. 1.ix.1971, *Salubeni* 1707 (SRGH).
Mozambique. N: Maniamba, Mepoche, fl. & fr.14.ix.1934, *Torre* 589 (COI; LISC). MS: Chimoio, Garuso Mt., fl. 5.iii.1948, *Garcia* 526 (LISC).
  Also in tropical East Africa, and Angola to West Africa, Central America. Riparian woodland, thickets, grassland, edges rivers and river banks, sandy soil and wet places; 500–1525 m.
  The var. *curtissii* (House) Stearn is a much stouter plant than var. *ochacea*. It occurs in Uganda and is naturalised in Cuba, Panama and Jamaica.

40. **Ipomoea obscura** (L.) Ker-Gawl. in Bot. Reg. **3**: t. 239 (1817).—Choisy in DC., Prodr. **9**: 370 (1845).—Hall. f. in Engl., Bot. Jahrb. **18**: 140 (1893) & **28**: 38 (1899).—Britten in Trans. Linn. Soc. London, 2. Ser., **4**: 29 (1894).—Dammer in Engl., Pflanzenw. Ost-Afr. C: 332 (1895).—Baker & Rendle in F.T.A. **4, 2**: 164 (1905).—Schinz Pl. Menyhart. in Beitr. Kenntn. Unteren Sambesi: 435 (1905).—Gibbs in Journ. Linn. Soc., Bot. **37**: 456 (1906).—Eyles in Trans. Roy. Soc. S. Afr. **5**: 455 (1916).—Gomes e Sousa in Bol. Soc. Est. Moçambique 32: 86 (1936).—van Oostr. in Blumea **3**: 519 (1940); in Fl. Males., Ser. 1, **4**: 471, fig. 44 (1953).—Wild, Guide Fl. Victoria Falls: 156 (1953).—Dandy in F. W. Andr., Fl. Pl. Anglo-Egypt. Sudan **3**: 116 (1956).—Meeuse in Bothalia **6**: 746 (1958). —Verdc. in Kew Bull. **13**: 209 (1958); in F.T.E.A., Convolvulaceae: 116, fig. 24, 1 (1963).—Heine in F.W.T.A. ed. 2, **2**: 349 (1963).—Roessler in Merxm. Prodr. Fl. SW. Afr. **116**: 15 (1967).—Binns H.C.L.M.: 40 (1968).—Ross, Fl. Natal: 296 (1972). —Jacobsen in Kirkia **9**: 171 (1973).—Compton, Fl. Swaziland: 478 (1976). Type from Java.

Perennial herb, with a taproot not thicker than a finger. Stems several to many, slender, prostrate to twining, up to 3 m. long, pilose or glabrescent. Leaf lamina ovate, rarely linear-oblong, 2·5–8·5 × 0·4–7 cm., acuminate or apiculate at the apex, cordate at the base, sometimes sagittate, entire or slightly undulate and often ciliate along the margin, pubescent or glabrescent on both surfaces; petiole up to 11 cm. long, pubescent or glabrescent as in the stems. Inflorescences 1–several-flowered; peduncle slender, 1–8 cm. long, glabrous or shortly hairy; bracteoles minute, narrow, acute; pedicels 1–2 cm. long, sometimes minutely verrucose, shortly hairy or glabrous, at first erect but in fruit reflexed and thickened towards the apex. Sepals subequal, oavte, ovate-orbicular, ovate-lanceolate or lanceolate, 4–8 mm. long, acute or apiculate, glabrous to pilose, in fruit all somewhat accrescent, ultimately often spreading or reflexed. Corolla funnel-shaped, yellow, orange, cream or white, concolorous or with purple centre, 1·5–2·5 cm. long, glabrous or the midpetaline areas thinly hairy towards the apices. Capsule globose, tipped with persistent style base. Seeds ovoid, black, appressed pubescent or velvety.

Leaf lamina ovate-cordate ·   ·   ·   ·   ·   ·   ·   ·   ·   var. *obscura*
Leaf lamina linear-oblong, sometimes as narrow as 4 mm., sagittate at the base
                                                                var. *sagittifolia*

Var. **obscura.**—*Meeuse in Bothalia* **6**: 747 (1958).—Verdc. in F.T.E.A., Convolvulaceae: 116 (1963).—Binns, H.C.L.M.: 40 (1968).—Ross, Fl. Natal: 296 (1972).—Jacobsen in Kirkia **9**: 171 (1973).

Convolvulus obscurus L., Sp. Pl., ed. 2: 220 (1762). Type from Java.

Ipomoea luteola R. Br., Prodr.: 485 (1810). Type from Australia.

Ipomoea fragilis Choisy in DC., Prodr., **9**: 372 (1845).—Baker & Rendle in F.T.A. **4**, 2: 165 (1905).—Dandy in F.W. Andr., Fl. Anglo-Egypt. Sudan **3**: 116 (1956). Type from S. Africa.

Ipomoea longipes Engl., Bot. Jahrb. **10**: 246 (1888) *nom. illegit.* non Garcke (1849). Type from Namibia?.

Ipomoea demissa Hall. f. in Engl., Bot. Jahrb. **18**: 129 (1893); **28**: 38 (1899).—Dammer in Engl., Pflanzenw. Ost-Afr. **C**: 331 (1895). Type from Tanzania.

Ipomoea inconspicua Baker in Kew Bull. **1894**: 71 (1894). Type: Malawi, Nakulambe, *Buchanan* 1881 (K).

Ipomoea obscura (L.) Ker-Gawl. var. *abyssinica* Hall. f. in Engl., Bot. Jahrb. **28**: 39 (1899). Type from Ethiopia.

Ipomoea obscura (L.) Ker-Gawl. var. *indica* Hall. f. in Engl., Bot. Jahrb. **28**: 39 (1899). Type as for *I. obscura* var. *obscura*.

Ipomoea fragilis Choisy var. *pubescens* Hall. f. in Bull. Herb. Boiss. **7**: 51 (1899).—Hall. f. in Warb., Kunene-Samb.-Exped. Baum: 347 (1903).—Baker & Rendle in F.T.A. **4**. 2: 165 (1905).—Eyles in Trans. Roy. Soc. S. Afr. **5**: 454 (1916).—Suessenguth & Merxm. in Trans. Rhod. Sci. Ass. **43**: 40 (1951).—Wild, Guide Fl. Victoria Falls: 156 (1953). Type: Malawi, N. of Lake Malawi, lower Plateau, *Thomson* 1880 (K, holotype).

Ipomoea obscura (L.) Ker-Gawl. var. *fragilis* (Choisy) Meeuse in Dyer, Fl. Pl. Afr. **31**: pl. 1222 (1956); in Bothalia **6**: 747 (1958).—Macnae & Kalk, Nat. Hist. Inhaca Isl., Moçamb.: 152 (1958).—Binns, H.C.L.M.: 40 (1968).—Ross, Fl. Natal: 296 (1972).—Jacobsen, Check-List Fl. Lomagundi Distr. Rhod.: 171 (1973).

Ipomoea obscura (L.) Ker-Gawl. var. *demissa* (Hall. f.) Verdc. in Kew Bull. **33**: 165 (1978).

Botswana. N: Dindinga Isl., fl. & fr. 27.iii.1975, *Smith* 1314 (K; SRGH). SW: c. 72 km. N. of Kang, fl. & fr. 18.ii.1960, *Wild* 5047 (K; SRGH). SE: Kahia Bangwaketse Reserve, fl. & fr. 12.ii.1961, *Smithers* in GHS 124668 (SRGH). **Zambia**. B: Masese, fl. 8.ix.1969, *Mutimushi* 3580 (K). N: Mbala, Mpulungu, Lake Tanganyika, 792 m., fl. & fr. 22.x.1967, *Simon & Williamson* 1155 (K; LISC; SRGH). W: Ndola, fl. & fr. 2.ix.1954, *Fanshawe* 1519 (K; SRGH). C: N. of Lusaka on the Great North Rd., fl. & fr. 9.viii.1952, *Angus* 182 (FHO; K). E: c. 4·8 km. E. of Chipata, 1000 m., fl. & fr. 7.i.1959, *Robson* 1054 (BM; K; LISC; SRGH). S: Machili, fl. & fr. 9.iii.1961, *Fanshawe* 6405 (K; SRGH). **Zimbabwe**. N: Mazoe, 1220 m., fl. & fr. viii.1917, *Walters* 2308 (K; SRGH). W: Bulalima Mangwe, 1 km. W. of Plumtree, c. 1300 m., fl. & fr. 27.xi.1972, *Norrgrann* 287 (K; SRGH). C: Harare, Alexandra Park, Sandringham Drive, 1494 m., fl. & fr. 16.x.1968, *Biegel* 2640 (K; LISC). E: Mutare Commonage, fl. & fr. 14.x.1948, *Chase* 1576 (BM; SRGH). S: 48–97 km. S. of Masvingo, fl. & fr. 16.iv.1948, *Rodin* 4241 (K; PRE). **Malawi**. N: Mzimba Distr., Viphya Plateau, c. 60 km. SW. of Mzuzu, 1676 m., fl. & fr. 11.xi.1973, *Pawek* 7494 (K; SRGH). C: Lilongwe, 1100 m., fl. & fr. 29.iii.1970, *Brummitt & Little* 9504 (K). S: Blantyre Distr., 24 km. W. of Chileka Airport on Rd. to Mpatamanga, 450 m., fl. & fr. 24.ii.1970, *Brummitt* 8715 (K; SRGH). **Mozambique**. N: Mandimba, fl. & fr. 10.iv.1942, *Hornby* 4526 (K; LISC; MAL; PRE; SRGH; UPS). T: Between Marueira and Songo, 1 km. from Songo, fl. & fr. 17.iv.1972, *Macedo* 5202 (LISC; LMA; SRGH). MS: Manica Village, fl. & fr. 29.iii.1966, *Wild* 7565 (K; LISC; SRGH). GI: Limpopo, Chicualacuala, fl. & fr. 28.iii.1974, *Balsinhas* 2668 (LMA). M; Mamaacha, Changalane, Estatuene, 527 m., fl. & fr. 9.xi.1967, *Balsinhas* 1146 (COI; LMA).

Widely spread through tropical Africa to S. Africa, Madagascar, the Mascarene Isl., the Seychelles, also tropical Asia to Queensland and Fiji, China and Formosa. Woodland, bushland, savanna, grassland, sandy soils, cultivated ground and roadsides; 0–1750 m.

Var. **sagittifolia** Verdc. in Kew Bull. **13**: 210 (1958); **33**: 165 (1978). Type from Tanzania.

Zimbabwe. W: Bulawayo, Buena Vista Farm, fl. & fr. 26.x.1975, *Cross* 255 (K; SRGH). C: Charter, 1463 m., fl. & fr. xi.1961, *Davies* 2936 (SRGH). S: Bikita, fl. & fr. 25.x.1967, *Plowes* 2863 (K). **Malawi**. S: Mangochi Distr., hill 4 km. NE. of Mangochi, 515 m., fl. & fr. 24.ii.1979, *Brummitt & Patel* 15468 (K; MAL; SRGH).

Also in Tanzania. Scrubland, and roadside; 1433–1463 m.

41. **Ipomoea verrucisepala** Verdc. in Kirkia **6**: 118, photo 1 (1967). Type: Zimbabwe, Chipuriro, Umvukwes Range, Mpingi Pass, *Drummond* 6860 (K, holotype; SRGH, isotype).

Annual or perennial herb. Stems twining or prostrate, slender, glabrous or scarcely hairy principally at the nodes. Leaf lamina ovate or ovate triangular, 2·5–13 × 1·1–8·5 cm., acuminate, mucronulate at the apex, cordate at the base, sparsely appressed pubescent to glabrescent on both surfaces; petiole 0·6–3 cm. long, slender, pubescent to glabrescent. Flowers axillary, solitary or geminate; peduncle 0·4–2·5 cm. long, slender, pubescent or glabrescent; pedicels 1·8–3·5 cm. long, thicker than peduncles, glandulous-muriculate; bracts linear, 2–3 mm. long. Sepals ovate-elliptic, 6–8·5 mm. long, acute or obtuse, glabrous but verruculous-tuberculate. Corolla funnel-shaped, white, 1·8–3 cm. long, glabrous. Capsule globose, glabrous, tipped with persistent style base. Seeds ovoid appressed pubescent or velvety.

Zimbabwe. N: Chipuriro, Umvukwes Range, Mpingi Pass, fl. 17.iii.1960, *Drummond* 6860 (K, holotype; SRGH, isotype). E: Mutare, Murahwa's Hill Commonage, 1113 m., fl. 28.iii.1966, *Chase* 8413 (K; LISC; SRGH).
   Not known from elsewhere. Woodland and roadsides; 1113–1280 m.

42. **Ipomoea sepiaria** Roxb., Fl. Ind., ed. Carey & Wall. **2**: 90 (1824).—Verdc. in Kew Bull. **15**: 8 (1961); in F.T.E.A., Convolvulaceae: 117 (1963). Type from India.
   Convolvulus diversifolius Scumach. & Thonn., Beskr. Guin. Pl.: 94 (1827). Type from Ghana.
   Batatas abyssinica A. Rich., Tent. Fl. Abyss. **2**: 64 (1851), non *Ipomoea abyssinica* Schweinf. (1867). Type from Ethiopia.
   Ipomoea diversifolia (Schumach & Thonn.) F. Didr. in Kjoeb. Vidensk Meddel. **1854**: 221 (1854). Type as for *Convolvulus diversifolius*.
   Ipomoea britteniana Rendle in Journ. Bot. **34**: 38 (1896).—Hall. f. in Engl., Bot. Jahrb. **28**: 43 (1899).—Baker & Rendle in F.T.A. **4**, 2: 167 (1905). Type from Kenya.
   Ipomoea hellebarda Schweinf. ex Hall. f. in Engl., Bot. Jahrb. **18**: 142 (1893) in syn.; **28**: 43 (1849).—Hiern in Cat. Afr. Pl. Welw. **1**: 737 (1898) incl. var. *sarcopoda*.—Baker & Rendle in F.T.A. **4**, 2: 170 (1905).—Hutch & Dalz., F.W.T.A. **2**: 216 (1931). Type from Ethiopia.
   Ipomoea homblei De Wild. in Bull. Jard. Bot. Brux. 5: 38 (1915). Type from Zaire.
   Ipomoea maxima (L.f.) Sweet var. *sagittata* Verdc. in Kew Bull. **13**: 209 (1958), Type from Zanzibar.

Perennial or rarely annual. Stems several from a cylindrical woody tuberous rootstock, twining, yellow-brown, ridges, pilose to glabrous. Leaf lamina ovate-cordate, triangular, oblong-triangular, or lanceolate, 1–12·5 × 1–9·5 cm., acuminate or acute at the apex, truncate sagittate, hastate or with rounded lobes at the base, glabrous save for the margins which are minutely puberulous; petiole 1–4 cm. long. Inflorescences few–many flowered, rather dense; peduncle 1·5–18 cm. long, glabrous; bracts small, lanceolate to ovate, acute; pedicels 0·5–1·4 cm. long. Sepals elliptic-oblong or ovate, 4–8 mm. long, obtuse to acutish, glabrous, coriaceous with thinner margins, the outer sometimes verruculose. Corolla funnel-shaped, lilac-pink or almost white, sometimes with a purple or maroon centre, 2–6 cm. long; limb almost salver-shaped; tube narrow. Stamens and style included or very slightly exerted. Capsule globose, glabrous. Seeds subtrigonous, pale, c. 3 mm. in diam., densely tomentose.

Zimbabwe. E: Mutare, SW. of Murahwa's Hill, 1100 m., fl. & fr. 23.iii.1958, *Chase* 6860 (BM; K). **Malawi.** N: Karonga Distr., c. 32 km. W. of Karonga, on Stevenson Rd., 762 m., fl. & fr. 16.iv.1976, *Pawek* 11076 (K; SRGH). **Mozambique.** N: Mandimba, fl. & fr. 1.iv.1942, *Hornby* 4529 (K; PRE).
   In tropical east Africa, Zaire to west Africa and Angola, also through tropical Asia and Malaysia to Formosa and Australia (Queensland). Woodland, savanna and grassland; 760–1220 m.

43. **Ipomoea papilio** Hall. f. in Bull. Herb. Boiss. **6**: 543 (1898).—Rendle in Journ. Bot. **39**: 56 (1901).—Baker & Wright in Dyer, F. C. **4**: 63 (1904).—Bak. & Rendle in F.T.A. **4**, 2: 167 (1905).—Engl. in Sitz-Ber. Königl. Preuss. Akad. Wiss. Berl. **52**: 24 (1907).—Eyles in Trans. Roy. Soc. S. Afr. **5**: 455 (1916).—Suesseng. & Merxm. in Trans. Rhod. Ass. **43**: 41 (1951).—Meeuse in Bothalia **6**: 750 (1958).—Binns, H.C.L.M.: 40 (1968).—Jacobsen in Kirkia **9**: 171 (1973). Type from S. Africa (Transvaal).
   Ipomoea papilio Hall. f. forma *pluriflora* Merxm. in Mitt. Bot. Staatss. München **1**: 204 (1953). Type: Zimbabwe, Rusape, *Dehn* "S57" (M).

Perennial herb. Stems up to 3 m. long, slender, trailing or also climbing, puberulous when young, glabrous, pubescent or scabridulous later. Leaf lamina broadly cordate or cordate—reniform to cordate-ovate, 2–6·5 × 2–7 cm., usually abruptly acuminate, basal sinus always broad and rounded, entire or irregularly lobed or toothed in the lower half, glabrous or nearly so on both surfaces except for the minutely and obscurely ciliate margin; petiole 0·8–5 cm. Inflorescences 1–5–flowered; peduncle 1·5–9 cm., slender, hispidulous or scabrid as stems and petioles or pubescent; bracts ovate, minute, usually scabrid; pedicels somewhat thickening upwards, 5–12 mm. long, minutely hispidulous, scabrid or pubescent. Sepals unequal, thinly coriaceous, glabrous or minutely pubescent, accrescent; outer ones oblong, elliptic or ovate to somewhat spathulate, obtuse, 5–6 mm. long; inner ones considerably longer, obtuse to almost truncate or faintly emarginate, minutely mucronate, 7–9 mm. long. Corolla funnel-shaped with horizontally spreading limb, light magenta or purplish, 2–3·5 cm. long, glabrous, hardly lobed. Capsule globose or ovoid-conical, glabrous. Seeds brown, shortly pubescent with a dense tuft of white or yellowish short hairs around the hilum.

Botswana. SE: Mochudi, Phutodikobo Hill, 915–1067 m., fl. & fr. 17.iii.1967, *Mitchison 75* (K). **Zambia.** C: Mt. Makulu Research Stn., c. 19 km. S. of Lusaka, fl. & fr. 30.iv.1957, *Angus 1567a* (K; SRGH). S: Mazabuka, Mapanza, 1067 m., fl. & fr. 3.iv.1958, *Robinson 2824* (K; PRE; SRGH). **Zimbabwe.** N: Mazoe, Great Dyke, nr. Jonuella Farm, 1700 m., fl. & fr. iii.1974, *Goldsmith 6/74* (SRGH). W: Bulawayo, 1372 m., fl. & fr. 24.xi.1920, *Borle 3* (COI; K; PRE; SRGH). C: Gweru, 6 km. N. of Lalapanzi, 1200 m., fl. & fr. 29.i.1973, *Biegel 4207* (K; SRGH). E: Mutare, 1097 m., fl. & fr. 11.x.1951, *Chase 4137* (BM; SRGH). S: West Nicholson, fl. & fr. 20.iii.1956, *Munro s.n.* (K; PRE). **Malawi.** C: Dedza, fl. & fr. 8–9.x.1966, *Binns 383* (SRGH). **Mozambique.** T: Angónia, Ulongue, fl. & fr. 1.xii.1980, *Macuácua 1339* (LMA).

Also in S. Africa (Transvaal and Swaziland). Woodland, open bushland, grassland and roadside; 915–1375 m.

44. **Ipomoea lapathifolia** Hall. f. in Engl., Bot. Jahrb. **18**: 142 (1893).—Baker & Rendle in F.T.A. **4**, 2: 168 (1905).—Eyles in Trans. Roy. Soc. S. Afr. **5**: 454 (1916).—Wild, Guide Fl. Victoria Falls: 156 (1953).—Meeuse in Fl. Pl. Afr. **31**: t. 1209 (1956); in Bothalia **6**: 752 (1958).—Verdc. in Kew Bull. **15**: 8 (1961); in F.T.E.A., Convolvulaceae: 118 (1963). —Roessler in Merxm. Prodr. Fl. SW. Afr. **116**: 15 (1967).—Binns, H.C.L.M.: 40 (1968).—Compton, Fl. Swaziland: 477 (1976). Type: Mozambique, Quelimane, *Stuhlmann 109* (not tracable at HBG).

Perennial herb. Stems from a thin taproot, twining or prostrate, up to 3 m. long, slender, ridged, glabrous, puberulous or sometimes hirsute. Leaf lamina from broadly ovate to elliptic, oblong, lanceolate or linear 5–18 × 0·5–5·5 cm., obtuse or subacute at the apex, cuneate, rounded or truncate at the base, entire, coarsely sinuate-dentate or with small lobes above the base, herbaceous drying papery, green above, paler below, pubescent or glabrescent, rarely hairy; petiole 0·5–7 cm. long, minutely scabridulous to hirsute. Flowers several in close umbellate cymes; peduncle 3·5–27 cm. long; bracteoles triangular to lanceolate-subulate, erect, acute, 1–3 mm. long, early deciduous; pedicels up to 1 cm. long. Sepals unequal, subcoriaceous; outer ones triangular to oblong or lanceolate-oblong from a broad base, 6–7 mm. long, finely muriculate or verrucose on the back; oblong to ovate, 7–8 mm. long, with a rather broad hyaline membranous edge; all slightly accrescent and glabrous. Corolla hypocrateriform, white, often with a magenta or mauve centre, 2–5·5 cm. long, with well-defined midpetaline areas. Capsule globose, glabrous, apiculate. Seeds brown, glabrous or minutely puberulous.

Var. lapathifolia

Ipomoea zambesiaca Baker in Kew Bull. **1894**: 70 (Feb. 1894), non Britten (Mar. 1894).—Dammer in Engl., Pflanzenw. Ost-Afr. **C**: 333 (1895). Type: Mozambique, Shupanga and delta of the Zambezi R., *Kirk & L. Scott* (K, syntype).

Ipomoea hellebarda Hall. f. var. *lapathifolia* (Hall. f.) Hall. f. in Engl., Bot. Jahrb. **28**: 44 (1899).

Ipomoea dasyclada Pilger in Engl., Bot. Jahrb. **41**: 297 (1908). Type from Tanzania.

Ipomoea intricata Pilger in Engl., Bot. Jahrb. **45**: 221 (1910). Type from Tanzania.

Ipomoea lapathifolia Hall. f. var. *dasyclada* (Pilger) Verdc. in Kew Bull. **13**: 211 (1958).

Botswana. N: Ngamiland, fl. & fr. xii.1930, *Curson 410* (PRE). **Zambia.** N: Mporokoso, Kabwe Plain, Mweru-Wantipa, 100 m., fl. & fr. 15.xii.1960, *Richards 13720* (K). C: Lusaka, fl. & fr. 25.ii.1965, *Fanshawe 9223* (SRGH). E: nr. Chizombo, c. 8 km. S.of Mfuwe, c. 609 m., fl. & fr. 30.xii.1968, *Astle 5405* (K; SRGH). S: Muyuni, Mapanza, 1067 m., fl. & fr. 19.iii.1957, *Robinson 2157* (K; SRGH). **Zimbabwe.** N: Gokwe, Nhongo, c. 8 km. N. of Gokwe Rd., to Mahore/Copper Queen, fl. & fr. 6.iii.1964, *Bingham 1159* (K; SRGH). W: Nyamandhlovu Res. Station, fl. & fr. 20.iii.1962, *Denny 320* (SRGH), C: Gweru, 1402 m., fl. & fr. 12.xii.1966, *Biegel 1562* (SRGH). **Malawi.** N: Rumphi Distr., Livingstonia Escarpment, 1337 m., fl. & fr. 31.xii.1973, *Pawek 7676* (K; SRGH). C: Dedza, fl. & fr. 28.v.1966, *Agnew 297* (SRGH). S: 6 km. S. of Monkey Bay, 490 m., 28.ii.1970, *Brummitt 8796* (K; LISC: MAL). **Mozambique.** N: Malema, 16 km. from Entre-Rios to Ribaue, 650 m., fl. 3.iii.1964, *Torre & Paiva 10405* (LISC). Z: Chinde, Luabo, fl. & fr. 13.viii.1971, *Correira 130* (LMA). MS: Chimoio, Vila Machado, 4 km. of Muda, fl. & fr. 9.xi.1953, *Gomes & Pedro* 4635 (LMA). M: Namaacha, Between Changalane and Goba, fl. & fr. i.1950, *Pedro* 3843 (LMA).

Also in Uganda, Tanzania, Zaire and S. Africa. Woodland, savanna, grassland, swampy places, roadsides, ruderal; 100–1400 m.

The var. *bussei* (Pilger) Verdc. has usually ovate irregularly and deeply simuate-dentate leaves save at the apex and a corolla that is 2·5 cm long. It occurs in Tanzania.

45. **Ipomoea humidicola** Verdc. in Kew Bull. **33**: 166 (1978) Type: Zimbabwe, Hwange, *Rushworth* 1435 (K, isotype; SRGH, holotype).

Annual herb. Stems several, erect or decumbent, from a woody rootstock, rooting at the nodes, slender, glabrous. Leaf lamina lanceolate, 2–12 × 0·4–1·7 cm., entire, acute at the apex, hastate, sagittate, truncate or rotundate, glabrous; petiole 0·2–3 cm. long, slender, glabrous. Flowers axillary, solitary or geminate; bracts lanceolate, 2·5 mm. long, glabrous; pedicels 0·8–4 cm., long, slender, glabrous. Sepals unequal, ovate, 3·5–5 mm. × 2–5 mm., glabrous accrescent; inner ones wider, with hyaline membranous edge; outer ones verrucose-echinate. Corolla funnel-shaped, yellow, 8 mm. long. Capsule globose, glabrous. Seeds trigonous, black, densely griseo-pubescent.

Zambia. C: Luangwa Valley, S. Game Reserve, 14·5 km. E. of Long Dambo, 600 m., fl. & fr. 7.iii.1967, *Prince* 330 (K). **Zimbabwe**. W: Hwange National Park, 0·8 km. W. of Main Camp along Rd. to Dom, 1050 m., fl. & fr. 23.i.19068, *Rushworth* 1435 (SRGH, holotype; K, isotype). Also in Tanzania. Woodland, wet grassland and moist sands; 600–1050 m.

46. **Ipomoea welwitschii** Vatke ex Hall. f. in Engl., Bot. Jahrb. **18**: 146 (1893).—Hiern, Cat. Afr. Pl. Welw. **1**: 739 (1898).—Rendle in Journ. Bot. **39**: 57 (1901).—Bak. & Rendle in F.T.A. **4**, 2: 174 (1905).—Rendle in Journ. Bot. **46**: 181 (1908).—Rendle in Journ. Linn. Soc., Bot. **40**: 150 (1911).—Eyles in Trans. Roy. Soc. S. Afr. **5**: 456 (1916).—Brenan, T.T.C.L.: 172 (1949).—Meeuse in Bothalia **6**: 756 (1958).—Verdc. in Kew Bull. **13**: 211 (1958); in Kew Bull. **15**: 9 (1961); in F.T.E.A., Convolvulaceae: 119 (1963).—Roessler in Merxm. Prodr. Fl. SW. Afr. **116**: 17 (1967).—Binns, H.C.L.M. **39**: 40 (1968).—Jacobsen In Kirkia **9**: 171 (1973). Type from Namibia.

Ipomoea hystrix Hall. f. in Engl., Bot. Jahrb. **18**: 146 (1893). Type from Tanzania.
Ipomoea aspericaulis Baker in Kew Bull. **1894**: 70 (1894). Type from Angola.
Ipomoea welwitschii Hall. f. var. *latifolia* Britten in Journ. Bot. **32**: 85 (1894).—Hiern, Cat. Afr. Pl. Welw. **1**: 739 (1898).—Baker & Rendle in F.T.A. **4**, 2: 174 (1905). Type from Angola.
Ipomoea inamoena Pilger in Engl., Bot. Jahrb. **45**: 221 (1910). Type from Namibia.
Ipomoea semisecta Merxm. in Trans. Rhod. Sci. Ass. **43**: 41 (1951). Type: Zimbabwe, Marondera, *Dehn* 193a (M, holotype; BR, isotype).
Ipomoea multinervia Verdc. in Kew Bull. **13**: 212 (1958). Type: Zambia, Nsama to Mporokoso, *Bullock* 1372 (K, holotype).

Perennial herb with a woody rootstock or globose tuber as big as a cricket ball. Stems several, up to 50 cm. long, suberect or prostrate, sometimes rather stout, stiff, glabrous or minutely puberulous. Leaves often erect, glabrous or asperulous; leaf lamina linear-lanceolate, ovate-lanceolate to oblong, 4–22 × 0·3–4 cm., acute at the apex, usually tapering, rounded or cuneate at the base, occasionally incised or trisected, rigid, glabrous or with short bristles on the nerves and margins, often with distinct reticulate nervation; petiole stout, 5–10 mm., long, caniculate above, glabrous or minutely puberulous. Inflorescences axillary, 1–3-flowered, sometimes profusely flowered when the plant is almost leafless; peduncle up to 1 cm. long, stout, glabrous; bracts minute, lanceolate; pedicels subclavate up to 1·2 cm. long. Sepals subequal, lanceolate 1–1·4 cm. long, acuminate or subacute, glabrous or minutely puberulous. Corolla funnel-shaped, rose-pink or white with pink or mauve centre 5–9·5 cm. long, with distinct midpetaline areas ending in mucronate-aristate points. Capsule globose, coriaceous, apiculate, glabrous. Seeds brown shortly puberulous with a basal tuft of hairs near the hilum.

Botswana. SE: NW of Molepole, 1036 m., fl. & fr. 1.xii.1954, *Codd* 8922 (K; SRGH). **Zambia**. B: Mankoya, nr. Luena R., c. 14·5 km. ESE. of Mankoya, fl. & fr. 21.xi.1959, *Cookson* 6715 (SRGH). N: Mporokoso Distr., Mporokoso-Nsama Rd., Mweru-Wa-Ntipa, 1200 m., fl. & fr. 13.xii.1960, *Richards* 13661 (K). W: Solwezi to Kansanchi Mine, c. 8 km., fl. & fr. 14.ix.1952, *White* 3250 (FHO; K). E: Chadiza, 850 m., fl. & fr. 28.xi.1958, *Robson* 758 (K; LISC; SRGH). S: Namwala, Shakalongo Plain, Kafue National Park, fl. & fr. 5.xii.1962, *Mitchell* 15/45 (LISC; SRGH). **Zimbabwe**. N: Gokwe, c. 3·2 km. NW. of Gokwe, fl. & fr. 20.xi.1963, *Bingham* 968

(K; LISC; SRGH). W: Shangani, Gwampa Forest Reserve, 914 m., fl. i.1956, *Goldsmith* 80/56 (K; SRGH). C: Chimanimani, Dombotombo African Towship, fl. 1.i.1967, *Mavi* 113 (K; SRGH). E: Mutare, S. of Circular Drive, 1097 m., fl. & fr. 22.xii.1962, *Chase* 7919 (K; LISC; SRGH). S: Masvingo Distr., Great Zimbabwe National Park, fl. & fr. 10.xi.1970, *Chiparawasha* 141 (SRGH). **Malawi**. N: Mzimba Distr., 10 km. E. of Moambazi, 32 km. W. of main Rd., 1240 m., fl. & fr. 30.xii.1975, *Pawek* 10664 (K; SRGH). C: Kasungu National Park, 1036 m., fl. & fr. 6.xii.1970, *Hall-Martin* 1041 (SRGH). **Mozambique**. T: Moatize, nr. Zóbuè-Moatize Rd., 350 m., fl. & fr. 13.i.1966, *Correia* 430 (LISC).

Also in Sudan, Tanzania, Angola and Namibia. Woodland, savanna, grassland, sandy soil; 350–1680 m.

47. **Ipomoea barteri** Baker in Kew Bull. **1894**: 70 (1894).—Baker & Rendle in F.T.A. **4**, 2: 169 (1905).—Verdc. in F.T.E.A., Convolvulaceae: 119 (1963).—Heine in F.W.T.A. ed. 2, **2**: 350 (1963).—Binns, H.C.L.M.: 40 (1968).—Jacobsen in Kirkia **9**: 171 (1973). Type from Nigeria.

Perennial herb often tinged purplish, with a tuberous root-stock fusiform or globose, lactiferous. Stems very slender, twining or prostrate up to 1 m. long, often clothed with fine spreading hairs. Leaf lamina, linear, lanceolate to ovate, 3–10 × 0·15–4 cm., entire, membranous, acute or obtuse at the apex, cuneate or cordate at the base; glabrescent or hispidulous along the margins; petiole 2–15 mm. long, hairy. Flowers usually solitary, rarely paired, fading early in the morning; peduncle 0·3–2 cm., long, glabrescent; bracteoles up to 3 mm. long, subulate. Sepals ovate or orbicular, 6–8 mm. long, obtuse, with reddish papillae, more rarely almost smooth, hairy or glabrescent. Corolla funnel-shaped, crimson, rose or mauve or limb white and tube coloured, 5–7·5 cm. long; tube rather dilated in middle. Capsule globose, glabrous. Seeds black or pale brown, ovoid, shortly pubescent.

Leaf lamina with cuneate base · · · · · · · · · · var. *barteri*
Leaf lamina with cordate base · · · · · · · · · var. *cordifolia*

Var. **barteri**

Ipomoea hanningtonii Baker in Kew Bull. **1894**: 70 (1894). Type from Tanzania.
Ipomoea klotzschii Dammer in Pflanzenw. Ost.-Afr. **C**: 332 (1895). Type: Mozambique, Sena, *Peters* 8 (B, holotype †).
Ipomoea barteri var. *stenophylla* Hall. f. in Bull. Herb. Boiss. **6**: 543 (1898).—Baker & Rendle in F.T.A. **4**, 2: 169 (1905). Type as for *Ipomoea klotzschii* Dammer.

Zambia. W: Mwinilunga Distr., 24 km. W. of Mwinilunga, on Matonchi Rd., nr. Musangila R., 1350 m., fl. 22.i.1975, *Brummitt, Chisumpa & Polhill* 13965 (K; NDO) C: Kapiri Mposhi, fl. & fr. 22.i.1955, *Fanshawe* 1819 (K). E: Lundazi, fl. & fr. 7.ii.1968, *Anton-Smith* in GHS 201767 (SRGH). S: Mochipapa, Gwembe, 1220 m., fl. & fr. 10.iii.1962, *Astle* 1475 (K; SRGH). **Zimbabwe**. N: Urungwe, c. 6.4 km. N. of Chinoye, 610 m., fl. & fr. 1.iii.1958, *Phipps* 982 (K; SRGH).C: Harare, Prince Edward Dam, fl. & fr. 8.i.1952, *Wild* 3742 (K; SRGH). E: Mutare, N. of Barrydale, 1220 m., fl. & fr. 9.i.1955, *Chase* 5439 (BM; SRGH). **Malawi**. N: Mzimba Distr., Viphya Plateau, Vernal Pool, c. 60 km. SW. Mzuzu, 1676 m., fl. & fr. 17.ii.1973, *Pawek* 6454 (K; SRGH). C: Dedza, Chongoni Forest, fl. & fr. 11.ii.1969, *Salubeni* 1260 (K; SRGH). **Mozambique**. N: NW. of Mandimba, fl. 26.i.1942, *Hornby* 3531 (K). T: Tete, Macanga, Furancungo Mt., 1400 m., fl. & fr. 15.iii.1966, *Pereira & Al.* 1727 (LMU).

Also in Uganda, Kenya, Tanzania, Zaire extending to West Africa. Woodland, grassland and sandy ground; 600–1800 m.

Var. **cordifolia** Hall. f. in Bull. Herb. Boiss. **6**: 543 (1898).—Wingfield in Kew Bull. **32**: 799 (1978). Type from Angola.

Ipomoea humifera Rendle & Britten in Journ. Bot. **32**: 177 (1894).—Baker & Rendle in F.T.A. **4**, 2: 168 (1905). Type as for *Ipomoea barteri* Baker var. *cordifolia* Hall. f.

Zambia. W: Mwinilunga Distr., between R. Kamakonde and R. Kamulende, c. 6·4 km. SW. of Matonchi Farm, fl. & fr. 17.ii.1938, *Milne-Redhead* 4612 (BM; K).

Also in Tanzania and Angola. Woodland and grassland.

48. **Ipomoea richardsiae** Verdc. in Kirkia **6**: 117 (1967). Type: Zambia, Mwinilunga Distr., Kalenda Village, *Richards* 17292 (K, holotype).

Perennial herb with a woody vertical rootstock. Stems erect, slender, up to 45 cm. long, with several erect shoots tufted, somewhat zigzag at the apex. Leaf lamina very variable in shape and size, linear oblong, linear-lanceolate, up to largely lanceolate or oblong-lanceolate, 3–12·5 × 0·3–3 cm., largely acute at the apex, cuneate at the base, greyish-pubescent, more so beneath; petiole 1–3 mm. long. Inflorescences numerous, axillary, 1–4-flowered; peduncle 0–1·6 cm. long, greyish-pubescent; pedicels 4–9 mm. long; bracts 1·5–5·5 mm. long, subobtuse at the apex, mucronulate; outer ones greyish-pubescent, inner ones with hyaline margin. Corolla tubular–funnel-shaped, white tinged with mauve, 2·8–3·7 cm. long; tube cylindric; limb infundibuliform. Capsule globose, glabrous. Seeds minutely pubescent.

Zambia. W: Mwinilunga Distr., Kalenda Ridge, W. of Matonchi Farm, fl. & fr. 2.xi.1938, *Milne-Redhead* 3058 (BM; K; SRGH).
Not known elsewhere. Woodland, on rock outcrop; 1200–1300 m.

49. **Ipomoea aquatica** Forssk., Aegypt.–Arab.: 44 (1775).—Baker & Rendle in F.T.A. **4**, 2: 170 (1905).—Eyles in Trans. Roy. Soc. S. Afr. **5**: 453 (1916).—Fries, Wiss, Ergebn. Schwed. Rhod.—Kongo-Exped. **1**: 270 (1916).—van Ooststr. in Fl. Males., Ser. 1, **4**: 473, fig. 47–8 (1953).—Brenan in Mem. N. Y. Bot. Gard. **9**: 7 (1954).—Dandy in F. W. Andr., Fl. Pl. Anglo-Egypt. Sudan **3**: 121, fig. 33 (1956).—Meeuse in Bothalia **6**: 753 (1958). —Verdc. in F.T.E.A., Convolvulaceae: 120 (1963).—Heine in F.W.T.A. ed. 2, **2**: 349 (1963).—Roessler in Merxm. Prodr. Fl. SW. Afr. **116**: 13 (1967).—Binns, H.C.L.M.: 39 (1968).—Ross, Fl. Natal: 295 (1972).—Gibbs-Russell in Kirkia **10**: 490 (1977). Type from Yemen.
Ipomoea reptans sensu Roem. & Schultes, Syst. **4**: 244 (1819) et auctt. mult.—Hutch. & Dalz., F.W.T.A. **2**: 215 (1931).—Williams, Useful Ornam. Pl. Zanzibar, Pemba: 312 (1949) non *Convolvulus reptans* L.

Annual or perennial herb. Stems several from a stout woody base, prostrate or floating, thick, semi-succulent, 2–3 m. long, rooting at the nodes glabrous or hairy at the nodes. Leaf lamina very variable in shape and size, ovate, triangular, ovate-oblong, lanceolate or linear, 3–15 × 1–10 cm., acute, acuminate or rarely obtuse at the apex, truncate, cordate or rounded at the base or sagittate to hastate, entire or coarsely dentate; petiole 3–25 cm. long, thick, glabrous. Inflorescence 1–few-flowered; peduncle 1–14 cm. long, glabrous; pedicels 2–6·5 cm. long, glabrous; bracts minute, narrow, acute, 1·5–2 mm. long. Sepals ovate or ovate-elliptic 6–12 mm. long, blunt or more or less acute, sometimes more or less tuberculate, margins thin and pale. Corolla funnel-shaped with a narrow tube, purple or pink or white with a deeper centre, 4–10·5 cm. long. Capsule globose, glabrous. Seeds densely pubescent.

Caprivi Strip. E. bank of Okavango R., just N. of the Botswana border, 1030 m., fl. 27.iv.1975, *Gibbs Russell* 2819 (SRGH). **Botswana**. N: Dikgathong, Thagoe R., fl. 7.x.1934, *Smith* 1152 (K; SRGH). **Zambia**. B: Mongu Distr., Barotse flood plain, fl. 19.iii.1964, *Verboom* 1201 (K). W: Kafubu R., Ndola, fl. & fr. 12.ii.1963, *Fanshawe* 7653 (SRGH). C: Mpika Distr., Luangwa Game Reserve, 27.iv.1965, *Mitchell* 2400 (K; SRGH). S: Machili, fl. 22.ii.1961, *Fanshawe* 6316 (K; SRGH). **Zimbabwe**. N: Urungwe, Mana Pools, fl. & fr. iii.1971, *Guy* 1618 (K; SRGH). W: Hwange, Kaungula, 914 m., fl. iv.1955, *Davies* 1132 (SRGH). E: Chipinge, Dakati Pan, Chibuwe, Sabi Valley, 457 m., fl. & fr. 26.iv.1967, *Plowes* 2857 (LISC; SRGH). **Malawi**. N: Karonga Distr., between Karonga and Chilumba, 500 m., fl. 24.iv.1975, *Pawek* 9531 (K; SRGH). C: Salima, on main Blantyre Rd., S. of Salima, fl. 2.iii.1977, *Grosvenor & Renz* 1001 (K; SRGH). S: Chikwawa Distr., Lengwe Game Reserve, 107 m., fl. 2.vii.1970, *Hall-Martin* 802 (K; PRE). **Mozambique**. Z: Chinde, Luabo, fl. & fr. 6.iii.1971, *Correira* 126 (LMA). T: Mutarara, fl. & fr. 11.xi.1971, *Haffern* 68 (SRGH). MS: Gorongosa National Park, fl. & fr. 10.vi.1966, *Macedo* 2159 (LMA). GI: Chibuto, nr. Mondiane, fl. & fr. 19.vi.1960, *Lemos & Balsinhas* 147 (BM; COI; K; LISC; LMA; SRGH).
Tropical East Africa, West Africa, Angola and Namibia; pantropical. Swampy places, savanna, banks and margins of rivers; 0–1500 m.

50. **Ipomoea stolonifera** (Cyrillo) J. F. Gmel., Syst. Nat. ed. 13, **2**: 345 (1791).—Hiern, Cat. Afr. Pl. Welw. **1**: 738 (1898).—Baker & Rendle in F.T.A. 4, 2: 171 (1905).—van Ooststr. in Blumea **3**: 540 (1940); Fl. Malesiana **4**: 478, fig. 51–2 (1954).—Heine in F.W.T.A. ed. 2, **2**: 350 (1963). Type from Italy.
Convolvulus stoloniferus Cyrillo, Pl. Rar. Neap. **1**: 14, t.5 (1788).

Perennial glabrous herb from a stout tuberous root. Stems trailing widely on the sands of seashore or running just beneath the surface, sending up short erect leaf branches, rooting at the nodes, glabrous. Leaf lamina very variable in shape, thick, rather fleshy, often of various forms on the same plant, linear, lanceolate, ovate or oblong, entire or 3–5-lobed, 1·5–5 × 1–3 cm., obtuse, truncate or cordate at the base, obtuse or emarginate to 2-lobed at the apex; petiole 0·4–4 cm. long. Inflorescences 1–3 flowered; peduncle 10–15 mm. long; pedicels 8–15 mm. long, accrescent in fruit; bracts minute, linear, 2–3 mm. long. Sepals unequal, inner ones 10–15 mm. long, outer ones shorter, all oblong, acute or obtuse, mucronulate, glabrous, subcoriaceous. Corolla funnel-shaped, white or pale yellow with a purple centre, 3–5 cm. long, glabrous. Capsule globose, glabrous. Seeds short tomentose or with longer hairs along the edges.

Mozambique. MS: Cheringoma, Chinizuia R., fl. v.1973, *Tinley* 2879 (K; LISC; SRGH).
West tropical Africa, Angola; S. Europe, Mediterranean region; also in Melaysia, tropical countries of both hemispheres. In sandy shores.

51. **Ipomoea asarifolia** (Desr.) Roem. & Schult., Syst. Veg.: 251 (1819).—Hall. f. in Engl., Bot. Jahrb. **18**: 145 (1893).—Hiern, Cat. Afr. Pl. Welw. **1**: 738 (1898).—van Ooststr. Blumea **3**: 539 (1940); Fl. Males. **4**: 477 (1953).—Heine in F.W.T.A. ed. 2, **2**: 1348 (1963). Type from Senegal.
Convolvulus asarifolius Desr. in Lam., Encycl. Méth. **3**: 562 (1789).
Ipomoea repens Lam., Tabl. Encycl. **1**: 467 (1791) non Roth 1821.—Baker & Rendle in F.T.A. **4**, 2: 172 (1905).—Hutch. & Dalz., F.W.T.A. **2**: 215 (1931).— Dandy in F.W. Andr., Fl. Pl. Anglo-Egypt. Sudan **3**: 120, fig. 32 (1956). Type from India.

Perennial herb, much resembling *I. pes-caprae* (52). Stems prostrate or sometimes twining, thick, terete or angular. Leaf lamina circular to reniform shaped, 3·5–7 × 3·5–8·5 cm., rounded at the apex, sometimes emarginate, mucronulate, cordate at the base with rounded lobes, glabrous, subcoriaceous; petiole 3–8·5 cm. long, rather thick with a deep longitudinal groove above, smooth or minutely muricated inflorescences axillary, often together with an axillary leaf shoot, cymosely 1-few-flowered; peduncle 2–5 cm. long; bracts ovate, minute; pedicels 1·5–3 cm. long. Sepals unequal, all elliptic-oblong, obtuse, mucronulate; inner ones 8–11 mm. long; outer ones shorter, 5–8 mm. long, more or less muricate. Corolla funnel-shaped, red-purple, up to 6·5 cm. long, glabrous. Capsule globose, glabrous, as large as a pea.

Zambia. B: Sesheke, fl. & fr. ii.1911, *Gairdner* 517 (K). **Mozambique**. T: Cahobra Bassa, right bank of Zambeze R., fl. 28.xi.1973, *Correia & al.* 3951 (LISC).
Also in Cape Verde Islands, Senegal, Nigeria, Sudan and Angola; tropical Asia and America. Open bush, termite mounds.

52. **Ipomoea pes-caprae** (L.) R. Br. in Tuckey, Narrat. Exped. River Zaire: 477 (Mar. 1818).—Sweet, Hort. Suburb. Lond.: 35 (July 1818).—Roth, Nov. Pl. Spec.: 109 (1821).—Choisy in DC., Prodr. **9**: 349 (1845).—Peters, Reise Mossamb., Bot. **1**: 238 (1861).—Peter in Engl. & Prantl, Pflanzenfam. **4**, 3a: 31 (1891).—Hall. f. in Engl., Bot Jahrb. **18**: 145 (1893).—Dammer in Engl., Pflanzenw. Ost-Afr. **C**: 332 (1895).—Hall. f. in Bull. Herb. Boiss. **5**: 376 (1897).—Engl., Pflanzenew. Afr. **1**, 1: 415, fig. 353 (1910).—van Ooststr. in Blumea **3**: 532 (1940).—Brenan, T.T.C.L.: 170 (1949).—Williams, Usef. Orn. Pl. Zanzibar Pemba: 310 (1949).—van Ooststr. in Fl. Males., Ser. 1, **4**: 475 (1953).—Macnae & Kalk, Nat. Hist. Inhaca Is., Moçamb: 152 (1958).—Meeuse in Bothalia **6**: 754 (1958).—Verdc. in Kew Bull. **13**: 211 (1958); F.T.E.A., Convolvulaceae: 121 (1963).—Heine in F.W.T.A. ed. 2, **2**: 347 (1963).—Amico & Bavazzano in Erb. Trop. Firenze **6**: 278 (1968).—Binns, H.C.L.M.: 40 (1968).—Williamson, Useful Pl. Malawi: 70 (1972).—Amico & Bavazzano, Erb. Trop. Firenze **49**: 542 (1978).—Munday & Forbes in Journ. S. Afr. Bot. **45**: 9 (1979). Type from India.
Convolvulus pes-caprae L., Sp. Pl.: 159 (1753).
Ipomoea biloba Forssk., Fl. Aegypt.—Arab.: 44 (1775).—Hiern, Cat. Afr. Pl. Welw. **1**: 739 (1898).—Baker & Rendle in F.T.A. **4**, 2: 172 (1905). Type from Yemen.

Perennial glabrous herb, much resembling the species above. Stem from a thick woody base, hollow, creeping, 5–30 m., long, often forming tangled mats. Leaves often secund, held erect; leaf lamina suborbicular, obreniform, quadrangular or elliptic, 3–9·5 × 3–10·5 cm., entire, conspicuously emarginate at the apex, very rarely rounded, usually appearing deeply 2-lobed, truncate, cuneate or subcordate at the base, subcoriaceous; petiole purplish, 4–16 cm. long, with two glands at apex. Inflorescences 1–many-flowered; peduncle erect, 3–16 cm. long; pedicels 1·2–4·5 cm. long; bracts or bracteoles ovate-lanceolate, 3–3·5 mm. long. Sepals subequal or outer ones a little shorter, ovate to elliptic-ovate, 5–12 mm. long, obtuse and mucronulate, very concave. Corolla funnel-shaped, pink or red-purple with a dark centre, 3–5·5 cm. long. Capsule globular, glabrous. Seeds blackish-brown, tomentose-villous.

Subsp. **brasiliensis** (L.) van Oststr. in Blumea **3**: 533 (1940).—Verdc. in F.T.E.A., Convolvulaceae: 121 (1963). Heine in F.W.T.A. ed. 2, **2**: (1963). Type from Brazil.

Convolvulus brasiliensis L., Sp. Pl.: 159 (1753). Type as above.
Ipomoea maritima R. Br., Prodr.: 486 (1810). Type from Australia.
Convolvulus bilobatus Roxb., Fl. Ind., ed. Carey & Wall. **2**: 73 (1824). Type from India.
Batatas maritima (R. Br.) Boj., Hort. Maurit.: 225 (1837).
Ipomoea pes-caprae (L.) R. Br. forma *arenaria* Dammer in Engl., Pflanzenw. Ost-Afr.: 332 (1895). Type from Tanzania.
Ipomoea pes-caprae (L.) R. Br. var. *emarginata* Hall. f. in Bull. Soc. Roy. Bot. Belge **37**: 98 (1898). Type as for *Convolvulus brasiliensis* L.
Ipomoea bilosa sensu Baker & Rendle in F.T.A. **4**, 2: 172 (1905) non Forssk. sensu stricto.

Leaf lamina elliptic, oblong-quadrangular or suborbicular, usually deeply emarginate or not very deeply bilobed, rarely rounded.

Zambia. N: Mbala, Lake Tanganyika, fr. 7.vii.1957, *Savory* 218 (K; SRGH). C: Kabwe, fl. xi.1928, *van Hoepen* 1305 (PRE). **Malawi**. N: Karonga Distr., Sangilo Hills, 549 m., fl. 3.i.1973, *Pawek* 6317 (K; SRGH). C: Dowa, 427 m., fl. 25.vii.1951, *Chase* 3850 (BM; K; SRGH). S: Mangochi, Lake Malawi at Cape Maclear, fl. 24.iii.1974, *Patel* 90 (SRGH). **Mozambique**. N: Metangula, Lake Malawi, fl. x.1964, *Magalhães* 32 (COI). Z: Pebane, fl. & fr. 5.x.1946, *Gomes Pedro* 2111 (LMA). MS: Beira, Macuti beach, fl. iii.1970, *Biegel* 3997 (SRGH). GI: Vilanculos, Bazaruto I., fl. 20.x.1958, *Mogg* 28499 (LISC; PRE). M: Maputo, Ponta do Ouro beach, fl. & fr. 27.xii.1948, *Gomes e Sousa* 3915 (COI; K; PRE).

Also in tropical east Africa, Zaire, Angola to west Africa and S. Africa; pantropical. Sandy sea-shores and more rarely on inland lake-shores and by roadsides; 0–1160 m.

The subsp. *pes-caprae* with more deeply bilobed leaves, occurs in Arabia and tropical Asia.

53. **Ipomoea transvaalensis** Meeuse in Bothalia **6**: 748 (1958).—Verdc. in F.T.E.A., Convolvulaceae: 122 (1963).—Jacobsen in Kirkia **9**: 171 (1973). Type from S. Africa (Transvaal).

Ipomoea convolvuloides Hall. f. in Engl., Bot. Jahrb. **18**: 140 (1893) *nom. illegit.* non Schinz (1888).—Baker & Wright in F.C. **4**, 2: 60 (1904). Type as for *Ipomoea transvaalensis*.

Perennial herb forming annual stems from a long fusiform tuberous rootstock. Stems suberect or prostrate up to 1 m. long, pilose. Leaf lamina triangular, oblong or ovate, 1·5–8 × 1–4·5 cm.; obtuse, acute or acuminate at the apex, emarginate or cordate at the base, entire, densely pilose with rather long silvery hairs on both surfaces; petiole 0·3–3·5 cm. long. Flowers solitary or in 2-flowered cymes; peduncles 1·5–6·5 cm., hairy; pedicels 6–15 mm. long, thickened upwards, less hairy than the peduncles; bracts minute, lanceolate. Sepals subequal, herbaceous, oblong or lanceolate with hyaline edges, 7–8 mm. long, acute, usually thinly and softly hairy outside. Corolla funnel-shaped, magenta, pink or white with purple centre, 3·5–5·5 cm. long; flower-bud with small but highly characteristic tuft of white hairs at the apex which become 5 small tufts in the open corolla which is otherwise glabrous. Capsule subglobose or somewhat ovoid, glabrous. Seeds densely and shortly velutinous.

Subsp. **orientalis** Verdc. in Webbia **13**: 324 (1958); in F.T.E.A., Convolvulaceae: 122 (1963). Type from Kenya.

Similar to subsp. *transvaalensis* but with a finer indumentum. Leaf lamina more ovate, wider (up to 5 cm. wide), shortly acuminate at the apex, deeply cordate at the base. Corolla white with a purple tube and a widely expanded limb.

Zimbabwe. N: Lomagundi, fl. & fr. xii.1965, *Wild* 7490 (K; LISC; SRGH).

Also in Kenya, Somalia and Tanzania. Open woodland savanna, copper oxide soils; 730–1160 m.

The subsp. *transvaalensis* is confined to southern Africa.

54. **Ipomoea protea** Britten & Rendle in Journ. Bot. **32**: 176 (1894).—Hiern, Cat. Afr. Pl. Welw. **1**: 739 (1898).—Baker & Rendle in F.T.A. **4**, 2: 173 (1905).—Verdc. in Kirkia **6**: 119 (1967). Type from Angola.

Perennial herb. Stems long, prostrate, branching up to 1·5 m. or more, rugulose, more or less minutely hirsute in the younger portion. Leaf lamina from broadly oblong to circular oblong, 3·5–6·5 × 2·5–6 cm., rounded or very obtuse at the mucronulate apex, more or less cordate at the base, upper surface scabridulous to sparsely pilose, lower surface lighter, markedly reticulate hirsute on the veins; petiole up to 8 mm. Inflorescences generally few to many-flowered dichasial cymes, sometimes 1-flowered; peduncles up to 20 mm. long, hispidulous; pedicels very short, hispidulous as the peduncles; bracts minute, ovate. Sepals subequal, ovate, 10–18 mm. long, shortly acute; deep wine-red at the base, glabrous inside, minutely warted outside, margin especially of the inner ones pellucid. Corolla funnel-shaped, mauve with deep mauve throat up to 6 cm. long. Fruit unknown.

Zambia. N: Mbala Distr., Mbala-Kambole Rd., c. 19 km. from Kambole, 1650 m., fl. 30.i.1959, *Richards* 10784 (K; SRGH).
Also in Angola. Sandy soils on roadside; 1650 m.

55. **Ipomoea vernalis** R.E. Fr., Wiss, Ergebn. Schwed, Rhod.-Kongo-Exped. **1**: 271 (1916).
Type: Zambia, Bwana Mkubwa, *Fries* 472 and 472a (UPS, syntypes).

Shrubby from a thick woody rootstock, with trailing habit, glabrous or nearly so, often flowering when completely leafless. Stems erect, virgate, sometimes branching hirsute in the younger portion, glabrous later. Leaves deciduous; lamina narrowly to largely oblong, ovate or obovate, 3–7·5 × 0·5–2·5 cm., acute or rounded, minutely apiculate at the apex, rounded at the base, glabrous or more or less pubescent; petiole 1–4 mm. long. Flowers axillary, 1-several, shortly pedunculate or fasciculate on the short older shoots; peduncle 0–20 mm.; pedicels more or less 10 mm. long; bracteoles minute, lanceolate more or less 3 mm. long. Sepals unequal, chartaceous, more or less largely ovate, apiculate, 6–10 mm. long, accrescent in fruit, with hyaline edges, glabrous, outer ones shorter. Corolla funnel-shaped, pale mauve, deep reddish mauve inside tube, 4–7 cm. long, glabrous. Capsule globose, glabrous, chartaceous. Seeds covered with brownish silky hairs.

Zambia. B: Zambesi, Balovale, 1067 m., fl. & fr. 5.viii.1952, *Kilger* 200 (PRE). N: 18 km. NE. of Chiengi, fl. 13.x.1949, *Bullock* 1368 (K). W: Mwinilunga Distr., Mwanamitowa R., fl. & fr. 16.viii.1930, *Milne-Redhead* 920 (K). C: Chilanga Fish Farm, 1250 m., fl. & fr. 21.ix.1962, *LNHC* 144 (K; SRGH). S: Mumbwa, W. of Lutale R., fl. 25.iii.1964, *van Rensberg* 2870 (K; SRGH). **Mozambique.** GI: Govuro, nr. Matobe, fl. & fr. 2.ix.1944, *Mendonça* 1971 (LISC).
Not known from elsewhere. On grasses of woodland, roadsides and alluvial soils; 900–1300 m.

56. **Ipomoea bolusiana** Schinz in Verh. Bot. Ver. Brandenb. **30**: 271 (Sept. 1888).—Hall. f. in Engl., Bot. Jahrb. **18**: 147 (1893); in Bull. Herb. Boiss. **7**: 53 (1899).—Baker & Rendle in F.T.A. **4**, 2: 175 (1905).—N.E. Br. in Kew Bull. **1909**: 123 (1909).—Eyles in Trans. Roy. Soc. S. Afr. **5**: 453 (1916).—Meeuse in Bothalia **6**: 758 (1958).—Roessler in Merxm., Prodr. Fl. SW. Afr. **116**: 13 (1967).—Ross, Fl. Natal: 295 (1972).—Compton, Fl. Swaziland: 477 (1976).—Verdc. in Kew Bull. **33**: 166 (1978). TAB. **24**. Type from Namibia.
Ipomoea simplex Hook. in Bot. Mag. **72**: t. 4206 (1846) *nom. illegit.* non Thumb.-Hall. f. in Engl., Bot. Jahrb. **18**: 146 (1893). Type from S. Africa.
Ipomoea angustisecta Engl. in Engl., Bot. Jahrb. **10**: 245, t. 7, fig. A (Oct. 1888).—Baker & Wright in Dyer F. C. **4**, 2: 49 (1904). Type from S. Africa.
Ipomoea mesenterioides Hall. f. in Bull. Herb. Boiss. **6**: 544 (1898).—Baker & Wright in Dyer F. C. **4**, 2: 50 (1904). Type from S. Africa (Transvaal).
Ipomoea bolusiana Schinz var. *abbreviata* Hall. f. in Bull. Herb. Boiss. **7**: 54 (1899). Type as for *Ipomoea angustisecta* Engl.
Ipomoea bolusiana Schinz var. *elongata* Hall. f. in Bull. Herb. Boiss. **7**: 54 (1899). Type from Angola.
Ipomoea praetermissa Rendle in Journ. Bot. **39**: 56 (1901).— Baker & Wright in Dyer F. C. **4**, 2: 48 (1904). Type from S. Africa (Transvaal).
Ipomoea simplex Thunb. var. *obtusisepala* Rendle in Baker & Rendle in F.T.A. **4**, 2: 174 (1905). Type: Zimbabwe, Harare, *Rand* 272 (BM, holotype).
Ipomoea bolusiana Schinz var. *pinnatipartita* Verdc. in Kirkia **6**: 118 (1967). Type from S. Africa (Transvaal).

Perennial, glabrous from a thick tuberous, subglobose or somewhat fusiform rootstock, 5–13 mm. in diam. Stems one or several, erect or prostrate, woody at the base, glabrous, terete, usually slender. Leaf lamina palmately 3–9 sect, with very narrow, linear or filiform segments 20–70 × 0·5–4 (10) mm., sometimes somewhat pinnate to distinctly pinnate, if 3 terminal segments are partly fused to form a common rachis, or entire, linear, sometimes linear spathulate, 4–10 (15) × 0·2–0·4 (0·7) cm.; petioles of dissected leaves up to about 2 cm. long, of simple leaves sometimes inconspicuous

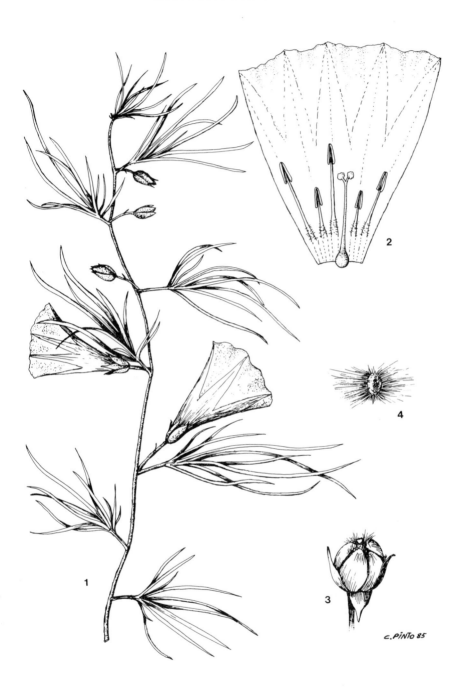

Tab. 24. IPOMOEA BOLUSIANA. 1, habit (×½); 2, corolla opened to show stamens and pistil (×1), 1–2 from *Lugard* 264; 3, fruit (×1); 4, seed (×1), 3–4 from *Hansen* 3275.

because the leaf is much narrowed at the base. Flowers axillary; peduncle very short, rarely up to 6 cm. long; pedicels very short, thickened, rarely up to 15 mm. long; bracteoles, small, lanceolate, often deciduous. Sepals ovate-lanceolate, lanceolate or oblong-lanceolate, sometimes ovate or elliptic, acute or acuminate, 8–16 mm. long, glabrous. Corolla funnel-shaped, magenta-pink, mauve or rosy-purple, darker in the centre, very rarely reported to be white, 4–7·5 cm. long and 4–6 cm. in diam. Capsule subglobose-conical, apiculate by the style-base, glabrous. Seeds covered with long shiny fawn hairs.

Botswana. N: Aha Hills, st. 13.iii.1965, *Wild & Drummond* 6962 (K; LISC). SW: c. 24 km. N. of Werda Police Station, fl. 23.ii.1960, *de Winter* 7496 (PRE). SE: Gaberone, Aedume Park, 1050 m., fr. 6.xi.1977, *Hansen* 3275 (K; SRGH). **Zambia**. B: Sesheke, fl. & fr. i.1925, *Borle* s.n. (PRE). N: Kawambwa, fl. & fr.16.xi.1957, *Fanshawe* 4047 (K). W: Mwinilunga Distr., S.W. of Matonchi Farm towards R. Ysongailu, fl. & fr. 14.xi.1937, *Milne-Redhead* 3226 (BM; K; LISC; SRGH). E: 8 km. S. of Chipata, 1067 m., fl. 17.xii.1959, *Wright* 24 (K). S: Namwala, Dambo, Kafue National Park, 1067 m., fl. & fr. 18.xii.1968, *Day* 17/69 (SRGH). **Zimbabwe**. N: Gokwe, Sengwa Res. Sta., 915 m., fl. & fr. 20.xii.1967, *Jacobsen* 3513 (PRE). W: Matobo, Matopos Res. Sta. 1341 m., fl. & fr. 21.iv.1975, *Dye* 32 (SRGH). C: Charter, 1219 m., fl. & fr. 27.ii.1926, *Eyles* 4594 (K; SRGH). E: Inyanga, 1700 m., fl. & fr. 24.ii.1930, *Fries, Norlindh & Weimarck* 3167 (PRE). S: Masvingo, nr. Sheppard's Hotel, fl. & fr. 18.xii.1962, *Grosvenor* 6 (SRGH). **Mozambique**. Z: Gurué, Mutuali, Comuè Mt., 600 m., fl. 10.xi.1967, *Torre & Correia* 16044 (LISC). T: Tete, nr. Boroma, fl. & fr. 21.ix.1942, *Mendonça* 340 (LISC). GI: Canicado, 37 km. from Massingir to Kruger Park frontier, fl. 13.xi.1970, *Correia* 1917 (LISC). M: Maputo, Boane, Porto Henrique Rd., 20–30 m., fl. 10.x.1980, *Shäfer* 7262 (K).
Also in Tanzania, Angola, Namibia, S. Africa (Swaziland, Natal) and Madagascar. Open woodland and savanna, grassland with scattered shrubs, rocky and sandy soils, 0–1465 m.

57. **Ipomoea fanshawei** Verdc. in Kirkia **6**: 119, phot. 3 (1967). TAB. **25**. Type: Zambia, Machili, *Fanshawe* 6001 (K, holotype; SRGH, isotype).

Perennial glabrous herb from a woody tuberous rootstock. Stems prostrate, slender, glutinous. Leaf-lamina rhomboid-lanceolate, 3–9 × 0·5–4 cm., acuminate or acute at the apex cuneate at the base, margin entire, sinuate or lobed, lobes linear or triangular, 3–20 × 1–2 mm., glabrous, thin, leathery, punctate, 6–8 palmatinervous, nerves prominent at the lower face; petiole 0·8–5·5 cm. long, canaliculate. Flowers axillary solitary; peduncle 3·7–5·5 cm. long; pedicels 0·5–1·2 cm. long slightly enlarged; bracts lanceolate, 5 mm. long. Sepals lanceolate, 1·5–1·8 cm. long, accrescent in fruit, acuminate, mucronulate, membranaceous, glabrous, glutinous. Corolla funnel-shaped, pale purple or pale mauve, 6 cm. long, glabrous. Capsule globose, coriaceous. Seeds ovoid, covered with shiny long fawn hairs.

Botswana. N: c. 13 km. N. of Francistown, 1097 m., fl. 14.i.1960, *Leach & Noel* 43 (SRGH). **Zambia**. S: Machili, fr. 8.iii.1960, *Fanshawe* 5417 (EA; K).
Not known from elsewhere. Woodland, dambo margins and open sandveld; 1097 m.

58. **Ipomoea tuberculata** Ker-Gawl. in Edwards, Bot. Reg. **1**: t. 86 (Feb. 1816).—Dandy in F. W. Andr., Fl. Pl. Anglo-Egypt. Sudan **3**: 118 (1956).—Verdc. in F.T.E.A., Convolvulaceae: 123 (1963).—Roessler in Merxm. Prodr. Fl. SW. Afr. **116**: 17 (1967). Type a plant cultivated in England from seed from Calcutta; no specimen now known.

Glabrous annual herb. Stems slender, climbing, smooth or tuberculate, 1–2·5 m. long. Leaf lamina circular in outline, pedately tripartite or digitately 5–9-lobed, up to 12 × 12 cm.; lobes linear-lanceolate to elliptic, acute, 3–8 × 1–3·5 cm.; petiole 2–7 cm. long; pseudo-stipules present. Inflorescences 1–3-flowered; peduncle 2–7 cm. long; pedicels thicker than the peduncle, clavate, up to 2·5 cm. long. Sepals orbicular to elliptic-ovate, obtuse, coriaceous, smooth or verruculose, somewhat unequal, the outer 3 shorter and broader, 6–12 mm. long, gibbous and 1–2-tuberculate at the base. Corolla funnel-shaped with a narrow tube, yellow or white with a purple centre, 5–10 cm. long. Capsule globose, coriaceous. Seeds subglobose-trigonous, brown with appressed pubescence and sometimes with long hairs in the angles.

Corolla mostly 4–5 cm. long · · · · · · · var. *tuberculata*
Corolla mostly rather large, up to 10 cm. long · · · · · · - var. *odontosepala*

Var. **tuberculata**
Ipomoea dasysperma Jacq., Eclog. Pl. **1**: 132, t. 89 (Aug. 1816).—Hall. f. Engl., Bot.

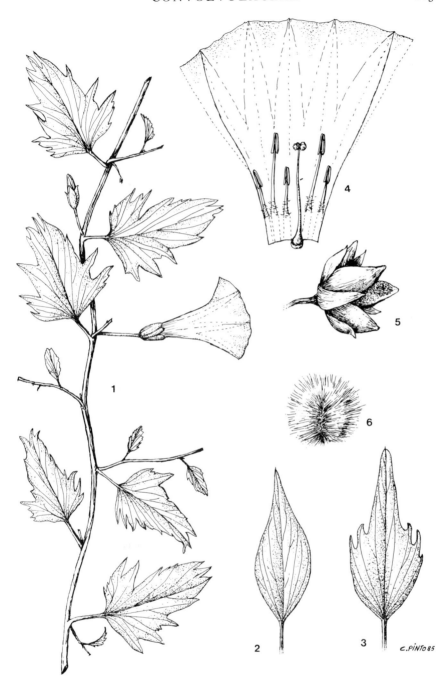

Tab. 25. IPOMOEA FANSHAWEI. 1, habit (×½), from *Leach & Noel* 43; 2 & 3 leaves (×½), from *Fanshawe* 5417; 4, corolla opened to show stamens and pistil (×1), from *Leach & Noel* 43; 5, fruit (×1); 6, seed (×1), 5–6 from *Fanshawe* 5417.

Jahrb. **18**: 148 (1893).—Baker & Rendle in F.T.A. **4**, 2: 179 (1905).—Meeuse in Bothalia **6**: 760 (1958). Type a plant cultivated at Vienna from seed stated to have come from China.
Ipomoea calcarata N.E. Br. in F.T.A. **4**, 2: 180 (1905); in Kew Bull. **1909**: 124 (1909). Type: Botswana, Ngamiland, Lugard 182 (K, lectotype).

Botswana. N: Botletle R., c. 20·9 km. from Maun, fl. iii.1968, *Lambrecht* 505 (K; SRGH). SE: nr. Zhilo Hill, nr. Shashi R., more or less 0·8 km. upstream from junction with Shashani R., fl. & fr. 2.v.1963, *Drummond* 7998 (SRGH). **Zambia**. S: Machili, fl. & fr. 2.iii.1961, *Fanshawe* 6353 (K; SRGH). **Zimbabwe**. N: Binga, Lusulu Veterinary Ranch, Busi R., fr. 24.ii.1965, *Bingham* 1415 (K; SRGH). W: Matobo, Champion Ranch, c. 8 km. WNW. Shashi R. and Shashani R. confluence, fl. & fr. 8.v.1963, *Drummond* 8187 (SRGH). E: Chipinge, Sabi Valley Exp. Stat., fl. & fr. x.1959, *Soane* 118 (SRGH). S: Mwenezi, nr. Malipati, fl. & fr. 24.iv.1961, *Rutherford-Smith* 685 (SRGH). **Mozambique**. T: Cahobra Bassa, more or less 550 m. from the dam, left edge of Zambezi R., 230–290 m., fl. & fr. 10.iv.1972, *Pereira & Correia* 1942 (LISC; LMU).

Also in Ethiopia, Namibia, S. Africa, extending to Sri Lanka and India; also reported from China, probably in error. Dambo margin woodland, edge of riverine forest, mixed scrub and river banks; 230–1600 m.

Var. **odontosepala** (Baker) Verdc. in Kew Bull. **14**: 341 (1960). Type from Tanzania.
Ipomoea odontosepala Baker in Kew Bull. **1894**: 73 (1894).—Baker & Rendle in F.T.A. **4**, 2: 180 (1905).
Ipomoea saccata Hall. f. in Engl., Bot. Jahrb. **28**: 48 (1899).—Baker & Rendle in F.T.A. **4**, 2: 180 (1905). Type from Tanzania.
Ipomoea dasysperma Jacq. var. *odontosepala* (Baker) Verdc. in Kew Bull. 13: 214 (1958).

Zambia. B: Masese, fr. 10.v.1961, *Fanshawe* 6555 (K; SRGH). Also in Ethiopia and Tanzania. Deciduous thicket and bushland.

59. **Ipomoea venosa** (Desr.) Roem. & Schult., Syst. Veg. **4**: 212 (1819).—Verdc. in Kew Bull. **13**: 213 (1958); in F.T.E.A., Convolvulaceae: 125 (1963); in Kirkia **6**: 120 (1967); in Kew Bull. **33**: 167 (1978). Type from the Mascarene Isl.
Convolvulus venosus Desr. in Lam., Encycl. Méth. Bot. **3**: 566 (1791).—Mordant Delaunay, Herb. Amat. **6**: t. 388 (1822).—Drapiez, Herb. Amat. **6**: t. 387 (1833).

Glabrous perennial herb. Stems twining or prostrate, smooth. Leaf-lamina circular in outline, compound divided into 5–7 leaflets or palmately 5–9-lobed to the base, 5–17 cm. long and wide; leaflets narrowly lanceolate 7–8·5 × 1–2·4 cm., acuminate and apiculate or obovate-elliptic, 3 × 1·4–1·6 cm., cuneate at the base and emarginate at the apex; lobes narrow lanceolate 5–8 × 0·1–0·5 cm.; outer lobes often bifid; petiole 1·5–5 cm. long; petiolules 0·2–1·0 cm. long. Inflorescences many-flowered cymes; peduncle 1–7·5 cm. long; pedicels short, up to 1·5 cm. long; bracts lanceolate up to 6 mm. long. Sepals ovate, 6–7·5 mm. long, acute. Corolla funnel-shaped, purple, violet or white, 4–7·5 cm. long. Capsule of subsp. *stellaris* (Baker) Verdc. var. *obtusifolia* Verdc. globose, glabrous, coriaceous and seeds trigonous, greyish, shortly pubescent, otherwise unknown.

Stems twining; leaf lamina compound, divided into 5–7 leaflets narrowly elliptic, 7–8·5 × 1–2·4 cm., acuminate and apiculate; corolla purple, violet or white   -   -   -   subsp. *venosa*
Stems twining or prostrate; leaf lamina palmately 5–9-lobed to the base or compound into 5 leaflets; lobes narrow lanceolate, 5–8 × 0·10–0·5 cm.; leaflets obovate-elliptic, 3 × 1·4–1·6 cm., cuneate at the base and emarginate at the apex; corolla purple or violet
                                                                                                 subsp. *stellaris*

Subsp. **venosa**.—Verdc. in Kirkia **6**: 120 (1967).
Ipomoea hornei Bak., Fl. Maurit. Seych.: 207 (1877). Type from Seychelles.
The characterization of this subsp. is sufficiently outlined in the key.

Zambia. B: c. 68 km. from Kaoma, on Mongu Rd., 25.xi.1970, *Anton-Smith* in GHS 211 783 (K; SRGH).
Also in the Seychelle Isl. On roadside, not common.

Subsp. **stellaris** (Baker) Verdc. in Kirkia **6**: 120 (1967).
Ipomoea stellaris Baker in Kew Bull **1894**: 73 (1894).—Dammer in Engl., Pflanzenw. Ost-Afr. **C**: 333 (1895).—Baker & Rendle in F.T.A. **4**, 2: 179 (1905).—Verdc. in F.T.E.A., Convolvulaceae: 125 (1963); in Kirkia **6**: 120 (1967). Type from Tanzania.
The characterization of this subsp. is sufficiently outlined in the key.

Stems prostrate; leaf lamina palmately 5–9-lobed to the base, lobes narrow lanceolate, 5–8 × 0·10–0·5 cm.   -   -   -   -   -   -   -   -   -   -   -   var. *stellaris*
Stems twining; leaf lamina compound into 5 leaflets obovate-elliptic, 3 × 1·4–1·6 cm., cuneate at base and emarginate at the apex   -   -   -   -   -   -   -   var. *obtusifolia*

Var. **stellaris**

Zambia. S: Bombwe, fl. 18.xii.1932, *Martin* 464 (FHO).
Also in Tanzania. Habit not known.

Var. **obtusifolia** Verdc. in Kirkia **6**: 120, phot. 4 (1967).

Mozambique. N: Maputo, between Maputo and Costa do Sol, fl. i.1950, *Pedro* 3844 (LMA,
holotype).
Not known elsewhere. Bushland and banks of river sands.

60. **Ipomoea cairica** (L.) Sweet, Hort. Brit.: 287 (1827).—Hall. f. in Engl., Bot. Jahrb. **18**:
(1893).—Dammer in Engl., Pflanzenw. Ost-Afr. **C**: 332 (1895).—van Ooststr. in Blumea
3: 542 (1940); in Fl. Malesiana, Ser. 1, **4**: 478 (1953).—Brenan in Mem. New York Bot.
Gard. **98**: 8 (1954).—Dandy in F. W. Andr., Fl. Anglo-Egypt. Sudan **3**: 119 (1956).
—Meeuse in Bothalia **6**: 761 (1958).—Verdc. in Kew Bull. **15**: 13 (1961); in F.T.E.A.,
Convolvulaceae: 125 (1963).—Heine in F.W.T.A. ed. 2, **2**: 351 (1963).—Binns,
H.C.L.M.: 40 (1968).—Ross, Fl. Natal: 295 (1972). TAB **26**. Type an illustration of
Convolvulus aegyptius in Vesling. Obs. in Prosp. Alp. Pl. Aegypt.: 75, fig. (1638)
(syn.).

Perennial from a tuberous rootstock, glabrous or nearly so. Stems twining or
prostrate, smooth or muriculate, up to 2 m. long, glabrous or villous at the nodes. Leaf
lamina ovate to circular in outline, palmately divided to the base into 5–7 lobes, 3 × 10
cm. long and wide; lobes lanceolate to ovate, elliptic or somewhat oblanceolate, acute or
obtuse and mucronulate at the apex, up to 5 × 1·6 cm.; outer lobes often bifid; petiole
2–6 cm. long, usually pseudostipulate by small leaves of developing or suppressed
axillary shoots; pseudo-stipules resembling the leaves but smaller. Inflorescences
lax, 1–many-flowered; peduncle 0·5–8 cm. long, branched; pedicels 1·2–3 cm.
long; bracteoles minute. Sepals subequal, ovate, 4–6·5 mm. long, obtuse to acute,
mucronulate, glabrous, sometimes verruculose, the edges membranous, outer ones
slightly shorter. Corolla broadly funnel-shaped, purple, red or white with purple
centre and purple tinge on outside of the limb or rarely entirely white, 4·5–6 cm. long,
(2·3–3 cm. long in var. *indica* Hall. f.). Capsule subglobose, glabrous. Seeds subglobose or ovoid, blackish, densely short-tomentose and also with long silky hairs
along the edges.

Var. **cairica**
Convolvulus cairicus L., Syst. Nat., ed. **10**: 922 (1759).
Ipomoea palmata Forssk., Fl. Aegypt.-Arab.: 43 (1775).—Baker & Rendle in F.T.A. **4**,
2: 178 (1905).—Schinz, Pl. Menyhart.: 435 (1905).—Fries, Wiss Ergebn. Schwed. Rhod.-
Kongo-Exped. **1**: 270 (1916).—H. Wild, Guide Fl. Victoria Falls: 156 (1953). Type from
Egypt.
Convolvulus tuberculata Desr. in Lam., Encycl. 3: 545 (1791). Type a plant cultivated in
Paris.
Ipomoea mendesii Welw., Ann. Conselho Ultramar Apont. Phyto-Geogr.: 584 (1859).
Type from Angola.

Botswana. SE: Gaberone Village, 975 m., fl. 8.x.1974, *Mott* 395 (SRGH). **Zambia**. N:
Mbala, Mpulungu, fl. 20.vi.1960, *Leach & Brunton* 10088 (K; SRGH). W: Mufulira, 1219 m., fl.
6.vi.1948, *Cruse* 364 (K). C: Lusaka, fl. & fr. 5.v.1957, *Noak* 226 (K; SRGH). S: Victoria Falls,
fl. & fr. 8.vii.1930, *Hutchinson & Gillett* 3435 (K). **Zimbabwe**. N: Shamva Mine, 3.viii.1927,
Young 100 (PRE). W: Victoria Falls Forest, 903 m., fl. & fr. iii. 1974, *Gonde* 81/74 (SRGH). C:
Harare 1463 m., fl. ix.1919, *Eyles* 1783 (K; PRE; SRGH). E: Mutare, fl. 7.ix.1948, *Chase* 874
(BM; SRGH). S: Masvingo 1909–12, *Monro* (BM). **Malawi**. N: Lake Nyassa, fl. 12.v.1952,
White 2826 (K; FHO). C: Nkhota Kota Distr., Benga, West Shore of Lake Nyassa, 470 m., fl. &
fr. 2.ix.1946, *Brass* 17495 (BM; K; SRGH). S: Blantyre Distr., just outside Blantyre City
boundary on road to Chileka Airport, fl. 30.vi.1970, *Brummitt* 11730 (LISC; MAL; SRGH).
Mozambique. Z: Chinde, between Chinde and Luabo, fl. & fr. 13.x.1941, *Torre* 3633 (BR; EA;
K; LISC; M; SRGH; WAG). T: Cahobra Bassa, Songo, c. 900 m., fl. & fr. 31.xii.1973, *Macedo*
5488 (LISC; LMU). GI: Vilanculos, S. Sebastião peninsula, 5 m., fl. 10.xi.1958, *Mogg* 29149
(LISC). M: Maputo, Polana, Beach, fl. & fr. 8.v.1964, *Balsinhas* 726 (LISC; LMA).
   Throughout tropical Africa. Also from the eastern Mediterranean region through Asia to
Formosa. Rainforest, bushland, swampy grassland, river edges, lake-shores, roadsides; 0–1676
m.
   The var. *indica* Hall. f. is rather intermediate between *I. cairica* (L.) Sweet and *I. hochstetteri*
House. It has a corolla about 2·3–3 cm.long and occurs in Kenya, Somalia and India.

61. **Ipomoea hochstetteri** House in Ann. New York Acad. Sci. **18**: 223 (1908) (sphalm
"*hochsteri*"); in Fedde, Repert. **8**: 231 (1910). Meeuse in Fl. Pl. Afr. **30**: t. 1189 (1955); in

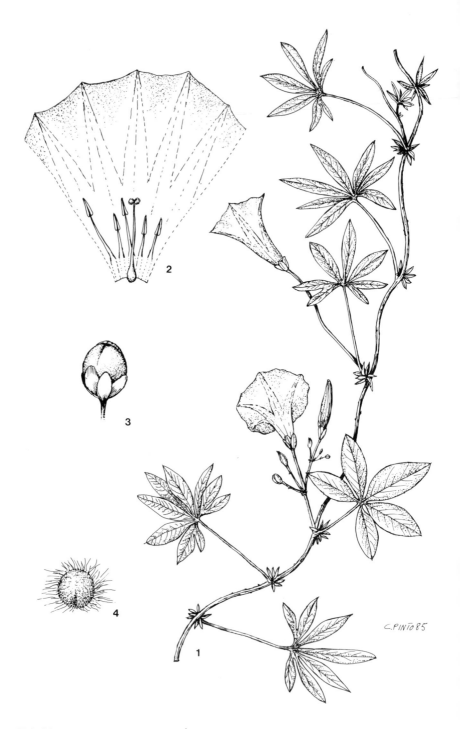

Tab. 26. IPOMOEA CAIRICA. 1, habit ($\times\frac{1}{2}$); 2, corolla opened to show stamens and pistil ($\times 1$), 1–2 from *Richards* 1050; 3, fruit ($\times 1\frac{1}{2}$), from *Sweet* 8; 4, seed ($\times 2$), from *Richards* 1050.

Bothalia **6**: 762 (1958).—Roessler in Merxm. Prodr. Merxm. SW. Afr. **116**: 14 (1967).
—Ross in Fl. Natal: 296 (1972).—Jacobsen in Kirkia **9**: 171 (1973). Type from Ethiopia.
Ipomoea quinquefolia Hochst. ex Hall. f. in Engl., Bot. Jahrb. **18**: 147 (1893) *nom. illegit.* non L. nec. Griseb. Type as above.
Ipomoea quinquefolia Hochst. ex Hall. f. var. *albiflora* Hall. f. in Bull. Herb. Boiss. **6**: 546 (1898). Type as above.
Ipomoea quinquefolia Hochst. ex Hall. f. var. *purpurea* Hall. f. in Bull. Herb. Boiss. **6**: 546 (1898). Types from Namibia.
Ipomoea kwebensis N.E. Br. in Kew Bull. **1909**: 123 (1909). Type as for *Ipomoea quinquefolia* Hochst. ex Hall. f. var. *purpurea* Hall. f.

Annual glabrous herb. Stems twining or prostrate, several metres long, with long internodes, smooth or finely striate, occasionally somewhat muriculate. Leaf lamina circular in outline, palmately divided to the base into 5 lobes, up to 13 × 12 cm. lobes elliptic, acute at both ends, with the two outer lobes usually deeply or completely redivided or rarely two lobes on the sides divided, 2·5–19·5 × 0·5–3 cm.; petiole smooth or slightly muriculate, up to 9 cm. long; pseudo-stipules present formed by young leaves of developing axillary shoots. Inflorescences cymosely 1–6-flowered; peduncle 0·6–9·5 cm. long, thick, smooth or slightly muriculate; pedicels stoutish, subclavate, 0·5–3 cm. long; bracteoles minute, subulate, deciduous. Sepals ovate, 5–7 mm. long, acute; mucronulate, with outer ones verruculose, often reflexed in fruit. Corolla funnel-shaped, white or mauve, 1·9–2·4 cm. long, the limb lobed and crinkly. Capsule globose, glabrous, enclosed by the persistent calyx. Seeds depressed-ovoid, dark brown, grooved on the outer face, densely silvery-grey tomentose and usually with long soft cottony hairs on the margins.

Botswana. N: Nata Area, at Nata R. delta, fl. & fr. 12.iv.1976, *Ngoni* 479 (K; SRGH). SW: 9 km. W. of Ghanzi, along Rd. to Mamumo, fl. & fr. 15.ii.1978, *Skarpe* 261 (K; SRGH). SE: Opapa, along airport Rd., fl. & fr. 19.iii.1974, *Allen* 29A (SRGH). **Zambia.** C: Lusaka, Stuart Park, 1219 m., fl. & fr. 7.v.1961, *Lusaka Natural History Club* 36 (K; SRGH). S: Monze, Lochinvar Ranch, *van Rensberg* 2098 (K; SRGH). **Zimbabwe.** N: Binga, Lusulu Veterinary Ranch, fl. & fr. 24.ii.1965, *Bingham*1403 (K; SRGH). W: Hwange, c. 8 km. from Main Camp along Shapi Rd., Hwange National Park, 1036 m., fl. & fr. 28.ii.1967, *Rushworth* 287 (SRGH). C: Chegutu, Poole Farm, fl. & fr. 3.iii.1948, *Hornby* 3137 (SRGH). E: Chipinge, Sabi Valley Experimental Station, fl. & fr. ix.1959, *Soane* 41 (SRGH). S: Gwanda, nr. Chiturupadzi store, c. 40 km. N-NW of the Bubye-Limpopo confluence, fr. 12.v.1958, *Drummond* 5767 (K; SRGH). **Malawi.** N: Rumphi Distr., c. 16 km. W. of Rumphi, 1067 m., fl. & fr. 24.iv.1977, *Pawek* 12614 (SRGH).
Also in Ethiopia, Uganda, Kenya, Tanzania, S. Africa (Natal). Namibia and also India (fide Meeuse). Sandy soils and roadsides; 930–1372 m.

62. **Ipomoea tenuipes** Verdc. in Kew Bull. **15**: 12 (1961) & in F.T.E.A., Convolvulaceae: 127 (1963).—Roessler in Merxm., Prodr. SW. Afr. **116**: 17 (1967). Type from Tanzania.
Convolvulus heptaphyllus Rottl. & Willd. in Ges. Naturf. Fr. Neue Schr. **4**: 196 (1803), pro parte (epithet not available in *Ipomoea*). Type from India.
Ipomoea radicans Choisy in DC., Prodr. **9**: 387 (1845) non Bl. (1826) *nom. illegit.* Type from Jamaica.
Ipomoea palmata Forssk var. *gracillima* Coll. & Hemsl. in Journ. Linn. Soc. Lond., Bot. **28**: 96 (1890). Type from Upper Burma.
Ipomoea gracilima (Coll. & Hemsl.) Prain in Journ. Asiatic. Soc. Bengal **63**: 111 (1894) non Peter (1891) *nom. illegit.*
Ipomoea heptaphylla sensu Meeuse in Bothalia **6**: 764 (1958) non (Roxb.) Voigt.

Annual or perennial glabrous herb. Stems very slender, twining or sometimes prostrate. Leaf-lamina circular in outline, palmately divided to the base, up to 5 cm. in diam., often much less; lobes elliptic to lanceolate 2·6–4 × 1 cm. usually obtuse and apiculate at the apex and narrowed and subpetiolued at the base, the outer ones often 2-lobed; petiole 4–5 cm. long; pseudostipules present resembling the leaves but much smaller. Flowers axillary, solitary or in 2–3-flowered inflorescences; peduncle 4–5 cm. long, very slender, almost filiform; pedicels slender but thicker than the peduncle, more or less 1·7 cm. long, somewhat subclavate; bracteoles minute. Sepals elliptic or almost orbicular, 3–5 mm. long, subequal with membranous edges, usually obtuse but mucronulate, often muriculate. Corolla funnel-shaped, rose-purple or dark mauve, 8–17 mm. long. Capsule globose, glabrous. Seeds subglobose, brown, velvety-pubescent, with 5–10 mm. long white cottony hairs at the edges.

Botswana. N: Nata R., c. 11 km. from mouth nr. Madsiara drift, 896 m., fl. & fr. 27.iv.1957, *Drummond & Brewer* 5262 (SRGH). SE: 60 km. NW. of Serowe, fl. & fr. 25.iii.1965, *Wild & Drummond* 7294 (K; SRGH). **Zambia**. C: Luangwa, Wafa, fl. & fr. 8.v.1965, *Mitchell* 2901 (K; SRGH). S: Gwembe Valley, fr. 7.vi.1963, *Lawton* 1081 (K). **Zimbabwe**. N: Kariba, 457 m., fl. & fr. vi. 1960, *Goldsmith* 89/60 (K; LISC; SRGH). W: Hwange, Deka R., fl. & fr. 21.vi.1934, *Eyles* 7968 (BM; K: SRGH). C: Chegutu, Poole Farm, fl. & fr. 3.iii.1948, *Hornby* 3137 (K). S: Gwanda, Umzingwane R., 609 m., fl. & fr. v.1955, *Davies* 1360 (K; SRGH). **Mozambique**. T: Mágoè, 37 km. from Chicoa to Mágoè, 300 m., fl. & fr. 16.ii.1970, *Torre & Correia* 18003 (LISC). GI: Guija, 2 km. after Caniçado, Souzuanine Rd., fl. & fr. 11.vi.1960, *Lemos & Balsinhas* 74 (BM; COI; K; LISC; LMA; SRGH). M: Maputo, Boane, Matutuine–Porto Henrique Rd., 30 m., fl. & fr. 8.viii.1980, *Schäfer* 7224 (K; LMU).

Also in Sudan, Tanzania, Namibia, Angola S. Africa (Transvaal), India and possibly West Indies. Riverine forest, savanna, grassland, river banks and cultivated ground; 30–900 m.

63. **Ipomoea coptica** (L.) Roth ex Roem. & Schult., Syst. Veg. **4**: 208 (1819); in Roth Nov. Pl. Spec.: 110 (1821).—Choisy in DC., Prodr. **9**: 384 (1845).—Peters, Reise Mossamb., Bot. **1**: 239 (1861).—Hall. f. in Engl., Bot. Jahrb. **18**: 147 (1893) & **28**: 45 (1899).—Dammer in Engl. Pflanzenw. Ost-Afr. **C**: 332 (1895).—van Ooststr. in Blumea **3**: 544 (1940) & in Fl. Males., Ser. 1, **4**: 479 (1953).—Dandy in F. W. Andr. Fl. Pl. Anglo-Egypt. Sudan **3**: 115 (1956).—Meeuse in Fl. Pl. Afr. 31: 1, t. 1217a (1956); in Bothalia **6**: 760 (1958).—Heine in F.W.T.A. Ed. 2, **2**: 350 (1963).—Verdc. in F.T.E.A., Convolvulaceae: 128 (1963). —Roessler in Merxm., Prodr. Fl. SW. Afr. the **116**: 14 (1967).—Binns, HC.L.M.: 40 (1968).—Ross, Fl. Natal: 295 (1972). Type from the Orient.

Annual, glabrous except for the base of the stamens. Stems several from the base, prostrate or twining, slender, up to 1·65 m. long, 4-angled, longitudinally striate and often finely muriculate especially on the angles. Leaf lamina orbicular in outline, digitately 5-lobed, 1–7 × 1–7 cm.; lobes linear to ovate or elliptic, dentate to deeply and coarsely once or twice pinnatifid, up to 6 × 2 cm., two outer lobes often bifid and the middle one longest, all lobes acute or subacute at the apex; petiole 2·5–20 mm. long; pseudo-stipules present, resembling leaves, but smaller, up to 2 cm. in diam. Inflorescences axillary, 1–3-flowered; peduncle 1–4·5 cm. long, narrowly alate; pedicels 4–10 mm. long, at first erect, in fruit bent downwards and thickening upwards; bracts lanceolate, 1·5–15 mm. long, acute, entire or laciniate. Sepals subequal, oblong or elliptic, 4–5 mm. long, accrescent in fruit, obtuse or cuspidate, muriculate, echinate or with undulating crests. Corolla funnel-shaped, white, pink or white with purple throat, about 12 mm. long. Capsule depressed-globose, minutely apiculate, 3-locular, glabrous. Seeds 6 or less by abortion, triquetrous, densely greyish-tomentose.

Main leaf lobes elliptic or ovate, deeply divided or toothed; bracts small 1·5–3 mm. long
                                                     var. *coptica*
Leaf segments linear, finely toothed, teeth narrow and very acute; bracts large, 1·5 cm. long, pinnatifid - - - - - - - - - - - - - - - - - - var. *acuta*

Var. **coptica**

Convolvulus copticus L. Mant. **2**: 559 (1771).
Convolvulus stipulatus Desr. in Lam., Encycl. **3**: 546 (1791). Type from Senegal.
Ipomoea dissecta Willd., Phytogr.: 5, t. 2 (1794).—Baker & Wright in Dyer F.C. **4**: 67 (1904).—Baker & Rendle in F.T.A. **4**, 2: 176 (1905).—Schinz in Pl. Menyhart. Beitr. Kennt. Unter. Sambesi: 435 (1905).—Hutch. & Dalz., F.W.T.A. **2**: 218 (1931).—Gomes e Sousa in Bol. Soc. Est. Moçamb. 32: 86 (1936).— Wild, Guide Fl. Victoria Falls: 155 (1953). Type from Guinea.
Convolvulus thonningii Schumach & Thonn., Beskr. Guin. Pl.: 98 (1827). Type from Ghana.
Ipomoea multisecta Welw. Ann. Conselho Ultramar Apont. Phyto-Geogr.: 589 (1859). Type from Angola.

Caprivi Strip. Mpilila Isl., 914 m., fl. & fr. 18.i.1959, *Killick & Leistner* 3335 (SRGH). **Botswana**. N: 26 km. SW of Maun, Moshu Bridge, 900 m., fl. & fr. 24.i.1972, *Biegel & Russell* 3781 (K; LISC; SRGH). SE: W. of Gaberone Dam, 975 m., fl. & fr. 3.iv.1976, *Mott* 945 (K; SRGH). **Zambia**. B: Masese, fl. & fr. 14.i.1961, *Fanshawe* 6137 (SRGH). N: Mporokoso, nr. Muzombwe, Mweru-Wa-Ntipa, 1036 m., fl. & fr. 15.iv.1961, *Vesey-FitzGerald* 3196 (K; LISC; SRGH). C: Lusaka Distr., Chinyunyu Hot Springs, 85 km. from Lusaka on Petauke Rd. fl. & fr. 12.ii.1975, *Brummitt & Lewis* 14318 (K; NDO; SRGH). S: Choma, 1311 m., fl. & fr. 28.iii.1955, *Robinson* 1207 (K; SRGH). **Zimbabwe**. N: Gokwe Distr., Sengwa Research Station, Ntaba-Mangwe Rd., 914 m., fl. & fr. 3.iv.1969, *Jacobsen* 529 (PRE; SRGH). W: Hwange, W. of Main Camp, Hwange National Park, 1067 m., fl. & fr. 14.ii.1969, *Rushworth* 1537 (K; SRGH). C: Chegutu, Poole Farm, 1219 m., fl. & fr. 25.ii.1948, *Hornby* 2900 (COI; K;

SRGH). E: Chipinge, Sabi Valley Experimental Station, fl. & fr. i.1960, *Soane* 239 (SRGH). S: Mwenezi, Malangwe R., SW. Mateke Hills, 624 m., fl. & fr. 6.v.1958, *Drummond* 5606 (K; LISC; SRGH). **Malawi**. C: Dedza, Ntakataka, fl. & fr. 10.iv.1969, *Salubeni* 1305 (K; SRGH). S: Farringdon Rd., nr. Mangochi, fl. & fr. 14.iii.1955, *Exell, Mendonça & Wild* 870 (BM; LISC). **Mozambique**. N: Amaramba, 10 km. from Nova Freixo to Mecanhelas, 600 m., fl. & fr. 14.ii.1964, *Torre & Paiva* 10572 (EA; K; LISC; PRE). T: Tete, Estima, fl. & fr. 22.iv.1972, *Macedo* 5234 (K; LISC; LMA; SRGH). MS: Chimoio, Vila Machado, fl. & fr. 15.ix.1965, *Balsinhas* 949 (COI; LMA). GI: Vilanculos, Mocoque, fl. & fr. 24.iii.1952, *Barbosa & Balsinhas* 4993 (BM; LMA). M: Maputo, Vila Luísa, Macaneta, fl. & fr. 1.iii.1971, *Balsinhas* 1788 (LMA).

Throughout tropical Africa, S. Africa, tropical Asia and northern Australia. Open mopane woodland, dry forest thicket, savanna, bushland, grassland, moist sands and cultivated fields; 90–1400 m.

Var. **acuta** Choisy in DC., Prodr. **9**: 384 (1845).—Baker & Rendle in F.T.A. **4**, 2: 177 (1905).—Verdc. in F.T.E.A., Convolvulaceae: 128 (1963). Type from Zanzibar.
  Ipomoea coptica sensu Roth, Nov. Pl. Sp.: 110 (1821) non (L.) Roem. & Schultes.
  Ipomoea palmatisecta Choisy in DC., Prodr. 9: 352 (1845). Type from Zanzibar.

Malawi. S: Zomba Distr., Lake Chilwa, fl. & fr. 2.v.1969, *Williams* 85 (SRGH).
Also in Kenya, Tanzania, Zanzibar Isl., Pemba Isl., and tropical Asia. On sandy soil by the coast.

64. **Ipomoea ticcopa** Verdc. in Bull. **13**: 213 (1958); in F.T.E.A., Convolvulaceae: 129 (1963). Type from Tanzania.
  Ipomoea coptica (L.) Roem. & Schult. var. *siphonantha* Hall. f. in Engl., Bot. Jahrb. **28**: 48 (1899). Type: Mozambique, Manica e Sofala, Lion's Creek, *Schlechter* 12218 (B, holotype; K, isotype).

Very similar to *Ipomoea coptica* (L.) Roth ex Roem. & Schult. in habit and foliage, probably directly derived from that. Leaf lamina circular in outline, palmately 5-lobed to the base, about 2·5 × 3·5 cm. in diam., lobes obovate, up to 2·5 × 2·2 cm., rounded in outline at the apex (much rounder than the last species), strongly cuneate at the base, sharply toothed, basal ones usually again bilobed; petiole 0·5–2·5 cm. pseudo-stipules present, resembling leaves, but much smaller. Inflorescences 1–2-flowered; peduncle 1·5–2 cm. long, often conspicuously winged; pedicels 5–7 mm. long; bracts leafy, up to 2·0 cm. long, with 3 pinnatifid lobes. Sepals elliptic, 6–7 mm. long, obtuse, the outer cuspidate, smooth, tuberculate or cristate. Corolla salver-shaped, white; tube narrowly cylindric 1·7–2·1 cm. long; limb 9 × 13 mm. Capsule globose, 3-locular. Seeds 6, trigonous, pubescent, brown.

Mozambique. MS: Lions-Creek, fl. 8.iv.1898, *Schlechter* 12218 (B; K).
Also in Kenya and Tanzania. Grassland, 30 m.

65. **Ipomoea trinervia** Schulze-Menz. in Notizbl. Bot. Gart. Berlin **14**: 109 (1938).—Verdc. in Kew Bull. **15**: 14 (1961) & in F.T.E.A., Convolvulaceae: 129 (1963). Type from Tanzania.

Annual herb. Stems erect or decumbent, pubescent. Leaf lamina obovate or spathulate up to 7 × 6 cm., pubescent, with apical part elliptic to orbicular up to 5 × 6 cm., acute at the apex, obscurely crenulate and a basal part narrowly cuneate, 2 × 0·7–1·5 cm.; petiole about 1 cm. long. Inflorescence an umbel-like cyme of more or less 6 flowers clustered; peduncle up to 2 cm. long, often much shorter, pubescent; pedicels 6 mm. long, pubescent; bracts linear-oblong, rounded at the apex, 3 mm. long, more or less 3-nerved, pubescent. Sepals linear-oblong, 5·5–9 mm. long. emarginate, conspicuously 3-nerved, shortly pubescent. Corolla trumpet-shaped, white, 2 cm. long, with narrow tube. Ovary elongate-conic, shortly pubescent. Capsule elongate-ovoid tipped by the persistent style, coriaceous, slightly pubescent. Seeds black, somewhat obscurely tuberculate, sparsely pubescent with long hairs.

Malawi. N: Likoma Isl., Lake Malawi, fl. 28.vi.1900 *Johnson* (K). S: Mangochi, Namaso Bay, above Anglican Cottage, 1800 m., fl. 26.iii.1979, *Blackmore* 710 (K). **Mozambique**. N: Macondes, 37 km. from Mueda to Mocimboa do Rovuma, 800 m., fl. & fr. 15.iv.1964, *Torre & Paiva* 12014 (LISC; PRE). MS: Chimoio, Bandula Mt., fl. & fr. 28.iii.1948, *Garcia* 787 (C; LISC; LMA; WAG).
Also in Tanzania. Open forest and rocky soil; 250–1.800 m.

66. **Ipomoea simonsiana** Rendle in Journ. Bot. **32**: 178 (1894).—Dammer in Engl., Pflanzenw. Ost-Afr. C: 333 (1895).—Baker & Rendle in F.T.A. **4**, 2: 168 (1905).—Verdc. in

F.T.E.A., Convolvulaceae: 129 (1963).—Binns, H.C.L.M.: 40 (1968). Type: Malawi, Lake Malawi, *Simons* (BM, holotype).

Annual herb branched near the base. Stems long, twining, slender, diffuse, obscurely pubescent. Leaf-lamina ovate in outline, 2·4–7·5 × 2·4–6·5 cm., acute, widely cordate at the base, margins with 2 or 3 deltoid acute or subacute lobes on each side of basal half, the whole leaf appearing more or less 5(7)-lobed, thin, very sparsely pubescent on both surfaces; petiole up to 5 cm. long, very slender. Flowers solitary or in 2-flowered cymes; peduncle 6–10 mm. long, slender; pedicels 5–12 mm. long, stouter than the peduncle; bracteoles minute, lanceolate. Sepals spathulate, more or less 7 mm. long, 3-nerved, with apical part ovate, apiculate, herbaceous, and the basal part linear-oblong, subcoriaceous. Corolla almost salver-shaped, white, 2·5 cm. long, lobes subacute, glabrous. Capsule globose, coriaceous, glabrous. Seeds brownish, densely pubescent with long hairs at the edges.

Zimbabwe. N: Chipuriro Distr., Angwa R., Mata Pools, fl. & fr. 3.vi.1965, *Bingham* 1542 (K; LISC; SRGH). W: Victoria Falls. fl. 1930, *Cheeseman* 218A (K). **Malawi**. N: Karonga Distr., c. 42 km. S. of Karonga, Ngala, 460 m., fl. & fr. 25.iv.1975, *Pawek* 9543 (K; SRGH). S: River Shire, 1863, *Kirk* (K). **Mozambique**. T: Tete-Songo Rd. Between Marueira and Chissua, 5 km. from Marueira, fl. & fr. 10.iv.1972, *Macedo* 5179 (LISC; LMA; LMU). MS: Chemba, Maringuè, 183 m., fl. & fr. 30.vi.1950, *Chase* 2466 (BM; K; SRGH).

Also in Tanzania. Savanna, thicket, grassland, lake shore and rocky soil; 183–550 m.

67. **Ipomoea alba** L., Sp. Pl.: 161 (1753).—Hall. f. in Meded. Rijksherb. Leiden **1**: 25 (1911) & 46: 19 (1922).—van Ooststr. in Blumea **3**: 547 (1940) & in Fl. Males. **4**: 480, fig. 53 (1953).—Meeuse in Bothalia **6**: 765 (1958).—Heine in F.W.T.A. ed. 2, **2**: 346 (1963). —Verdc. in F.T.E.A., Convolvulaceae: 130, fig. 22, 4 (1963).—Ross, Fl. Natal: 295 (1972). TAB. **27**. Type from India.

    Convolvulus aculeatus L., Sp. Pl.: 155 (1753). Type from tropical America.

    Ipomoea bona-nox L., Sp. Pl., ed. **2**: 228 (1762).—Williams, Useful Orn. Pl. Zanzibar and Pemba: 308 (1949). Type as for *Ipomoea alba* L.

    Calonyction speciosum Choisy in Mem. Soc. Phys. Genève **6**: 441, t. 1, fig. 4 excl. var. b (1834).—Hall. f. in Engl., Bot. Jahrb. **18**: 153 (1893).—Baker & Rendle in F.T.A. **4**, 2: 117 (1905). Type based on *Ipomoea bona-nox* L. (G, a specimen seen which substantiates this interpretation).

    Calonyction bona-nox (L.) Bojer, Hort. Maurit.: 227 (1837).—Hall. f. in Bull. Herb. Boiss. **5**: 1028 (1897). Type as for *Ipomoea bona-nox*.

    Calonyction aculeatum (L.) House in Bull. Torr. Bot. Club **31**: 590 (1904).—Hutch. & Dalz., F.W.T.A. **2**: 213 (1931).

Ornamental species, annual or perennial, glabrous or rarely pubescent. Stems twining or prostrate, thick, smooth or rarely muriculate. Leaf-lamina ovate or circular, in outline, entire or 3-lobed, 6–20 × 5–16 cm., acute, acuminate or obtuse at the mucronulate apex, cordate at the base; petiole 5–20 cm. long, slender. Inflorescence axillary, 1–several-flowered; peduncle stout, 1–24 cm. long; pedicels 0·7–1·5 cm. long, lengthening to 2·5–3 cm. long and becoming very thick and clavate in fruit. Sepals unequal, elliptic, coriaceous, often reflexed in fruit and glabrous; outer 2–3 sepals 5–12 mm. long with long awn-like appendage 4–10 mm. long at the apex; inner ones longer, 8–15 mm. long, shortly mucronulate. Corolla opening at night, scented, white or greenish-cream below; tube cylindrical 7–12 × 0·5 cm.; limb salver-shaped, 11–16 cm. wide. Capsule mucronulate 2·5–3 cm. long, glabrous. Seeds 4, ovoid, brown or black, glabrous and smooth.

Zimbabwe. E: Mutare, 1098 m., fl. 29.iv.1954, *Chase* 5225 (BM; K; SRGH). **Mozambique**. MS: Chimoio, Revuè R., nr. Sussendenga, 762 m., fl. & fr. 9.x.1969, *Leach & Cannell* 14410 (LISC; SRGH). M: Maputo, Chinhanguanine Rd., fr. 21.vii.1978, *Mafumo & Boane* 46 (LMU).

Also in Uganda, Kenya, Tanzania, S. Africa (Natal), cultivated in S. Africa and occasionally found as an escape; originally American but now pantropical as an escape and also cultivated as ornamental. Grassland, river banks, waste places and roadsides; 760–1158 m.

68. **Ipomoea turbinata** Lag., Gen. Sp. Pl.: 10 (1816).—Gunn in Brittonia **24**: 163, fig. 5 (1972).—Verdc. in Kew Bull. **33**: 167 (1978). Type from India.

    Convolvulus muriculatus L., Mant.: 44 (1767). Type a specimen of a plant cultivated at Uppsala from seed from Suratt, India (where it was probably also cultivated).

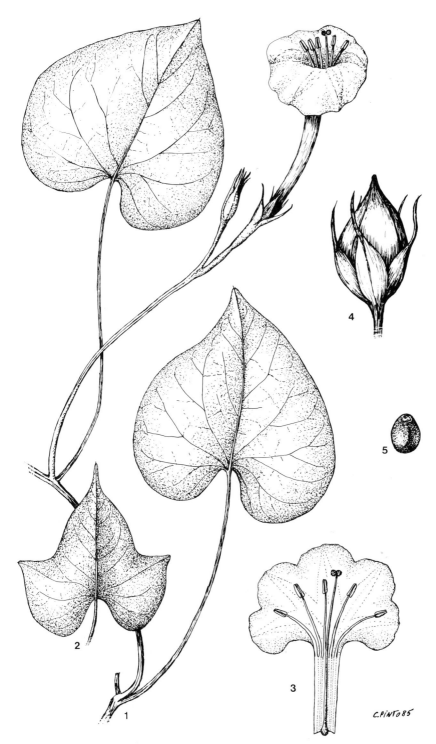

Tab. 27. IPOMOEA ALBA. 1, habit (×$\frac{1}{2}$), from *Biegel* 5225; 2, leaf (×$\frac{1}{2}$), from *Mafumo & Boane* 46; 3, corolla opened to show stamens and pistil (×$\frac{1}{2}$), from *Biegel* 5225; 4, fruit (×1), from *Mafumo & Boane* 46; 5, seed (×1), from *Chase* 5225.

Ipomoea muricata (L.) Jacq., Hort. Schoenbr. **3**: 40, t. 323 (1803) *nom. illegit.* non Cav. (1794).—Peter in Engl. & Prantl, Pflanzenfam. **4**, 3a: 29 (1891).—van Ooststr. in Blumea **3**: 551 (1940) & in Fl. Males., Ser. 1, **4**, 4: 481 (1963).—Verdc. in F.T.E.A., Convolvulaceae: 130 (1963).—Binns, H.C.L.M.: 40 (1968).
Calonyction muricatum (L.) G. Don., Gen. Syst. **4**: 264 (1837).—Hall. f. in Engl., Bot. Jahrb. **18**: 154 (1893); in Bull. Herb. Boiss. **5**: 1044 (1897).—Hiern in Cat. Afr. Pl. Welw. **1**: 742 (1898).—Baker & Rendle in F.T.A. **4**, 2: 118 (1905).—Hutch. & Dalz., F.W.T.A. **2**: 213 (1931).
Calonyction speciosum Choisy var. *muricatum* (L.) Choisy in DC., Prodr. **9**: 345 (1845) excl. syn. Willd.
Ipomoea shirensis Baker in Kew Bull. **1894**: 74 (1894) *nom. illegit.* non Oliv. (1884). Type: Malawi, Shire Highlands, *Kirk* (K, lectotype).
Ipomoea kirkiana Britten in Journ. Bot. **32**: 85 (1894). Type as for *Ipomoea shirensis* Baker.

Glabrous annual or perennial. Stems wide-climbing, muricate. Leaf lamina ovate or circular 7–16 × 6–14 cm., acuminate at the apex, cordate at the base, membranous; petiole 4–12 cm. long, smooth or muricate. Inflorescences 1–few-flowered lax dichasial or monochasial cyme; peduncle 3–10 cm. long, muricate; pedicels 1–2 cm. long, smooth, much thickened in the fruit; bracts minute. Sepals subequal, ovate oblong; outer sepals 6–7 mm. long with a long awn 4–6 mm. long; inner sepals 7–8 mm. long with a rather shorter awn. Corolla opening at night, pale bluish purple, 5–7·5 cm. long; tube narrow and cylindrical, 3–6 cm. long; limb funnel-shaped or salver-shaped. Capsule ovoid, 1·8–2 cm. tall. Seeds ovoid, 9–10 × 5 mm., flattened, black, glabrous.

Zambia. N: Mporokoso Distr., Mweru-Wa-Ntipa, 1050 m., fl. & fr. 10.iv.1957, *Richards* 9118 (K). S: nr. Mumbwa, 1911, *Macaulay* (K). **Zimbabwe**. N: Centenary Distr., Marabani T.T.L., fl. & fr. 11.iv.1965, *Bingham* 1456 (SRGH). W: Shangani, Lupani, cultivated in Harare, fl. & fr. 7.iv.1952, *Johnstone* s.n. (K; SRGH). **Malawi**. S: Shire Highlands, fl. & fr. xii.1881, *Buchanan* (K). **Mozambique**. T: Estima, Sanangoè R., fl. & fr. 22.iv.1972, *Macedo* 5238 (LISC; LMA).
Also in Gambia, Sudan, Angola and Tanzania. An American plant now widely distributed in the tropics of the Old World, often cultivated as ornamental. Naturalised in riverine forest, grassland, river margins; 1050 m.

69. **Ipomoea hederifolia** L., Syst. Nat., ed. **10**: 925 (1759).—Heine in F.W.T.A. ed. 2, **2**: 347 (1963).—Verdc. in F.T.E.A., Convolvulaceae: 132 (1963).—van Ooststr. in Fl. Java **2**: 491 (1965). TAB. **28**. Type from the West Indies.
Ipomoea angulata Lam., Illustr. **1**: 464 (1793).—van Ooststr. in Plumea **3**: 553 (1940) & Fl. Males., Ser. 1, **4**: 481, fig. 54 (1953). Type from Mauritius.
Ipomoea phoenicea Roxb., Fl. Ind., ed. Carey & Wall. **2**: 92 (1824). Type from India.
Quamoclit angulata (Lam.) Bojer, Hort. Maurit.: 224 (1837).
Quamoclit coccinea sensu Baker & Rendle in F.T.A. **4**, 2: 128 (1905).—Hutch. & Dalz., F.W.T.A. **2**: 213 (1931) non (L.) Moench.

Ornamental annual species. Stems twining, 2–5 m. long, terete or slightly angular, often contorted, glabrous or sparsely pilose. Leaf lamina ovate to circular in outline, 3–15 × 3–10 cm., acuminate and mucronulate at the apex, cordate at the base, the margin entire, angular, coarsely dentate or obscurely to deeply 3-lobed, mostly glabrous; petiole 3–12 cm. long, glabrous or sparsely pilose. Inflorescences lateral or terminal, few to several flowered, 10–35 cm. long, peduncle 3–20 cm. long, terete or angular, glabrous or pubescent; pedicels erect, 0·5–5 cm. long, enlarging in fruit; bracts minute, triangular, mucronulate. Sepals oblong, 2–3 mm. long with apex broadly obtuse or truncate, with an awn 2–4 mm. long. Corolla salver-shaped, scarlet glabrous; tube 2·8–4 cm. long, narrowed below, very slightly curved; limb 2–2·5 cm. in diam. Stamens and style exserted 8 mm. Capsule globose, glabrous. Seeds black, densely pubescent.

Zambia. W: Ndola, fl. & fr. 17.v.1954, *Fanshawe* 1206 (K; SRGH). **Zimbabwe**.: Victoria Falls, W. of the village, 880 m., fl. & fr. 2.v.1978, *Mshasha* 55 (SRGH). **Malawi**. N: Nkhata Bay Distr., c. 40 km. S. of N.B. junction, Chintece Beach, 472 m., fl. & fr. 23.v.1976, *Pawek* 11310 (SRGH). S: Mulanje, Likabula R., 680 m., fl. & fr. 6.v.1980, *Blackmore & Brummitt* 1470 (K). **Mozambique**. N: Lichinga, fl. & fr. 16.v.1948, *Pedro & Pedrogāo* 3597 (LMA; LMU).
Also in Uganda, Kenya, Tanzania; native of tropical America, now widely naturalised throughout the tropics. Brachystegia woodland, grassland, kalahari sand and lakeshore; 470–1220 m.

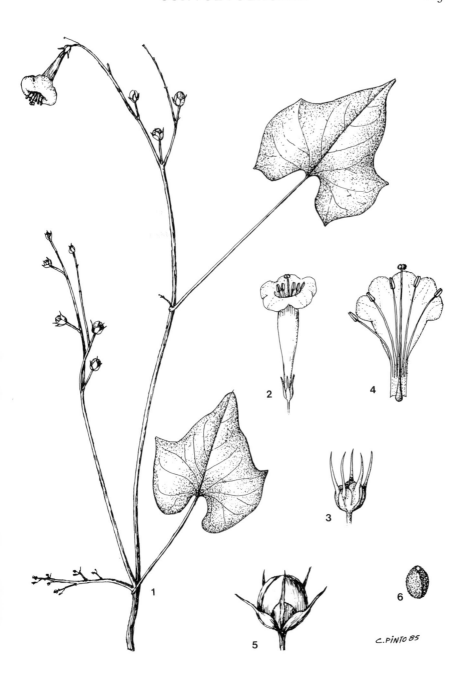

Tab. 28. IPOMOEA HEDERIFOLIA. 1, habit ($\times\frac{1}{2}$), from *Pawek 5115*; 2, flower ($\times$1); 3, calyx ($\times$2); 4, corolla opened to show stamens and pistil ($\times$1); 5, fruit ($\times$2), 2–6 from *Blackmore & Brummitt* 1470.

70. **Ipomoea shupangensis** Baker in Kew Bull. **1894**: 73 (1894).—Dammer in Engl., Pflan-
zenw. Ost-Afr. **C**: 333 (1895).—Baker & Rendle in F.T.A. **4**, 2: 170 (1905).—Engl. in
Sitz-Ber. Königl. Preuss. Akad. Wiss. Berl. **52**: 19 (1907).—Eyles in Trans. Roy. Soc. S.
Afr. **5**: 455 (1916).—Verdc. in F.T.E.A., Convolvulaceae: 132 (1963).—Drummond, List
Trees, etc. in Rhodesia: 271 (1975) TAB. **29**. Type: Mozambique, R. Zambezi, Shupanga,
Kirk 229, 230 (K, syntypes).

Ipomoea nuda Baker in Kew Bull. **1894**: 72 (1894).—Hiern, Cat. Afr. Pl. Welw. **1**: 741
(1898).—Baker & Rendle in F.T.A. **4**, 2: 169 (1905), non Peter (1891). Type from Angola.

Glabrous perennial liane. Stems ridged, up to 6 m. long, woody with a much-
wrinkled surface, often yellowish. Leaf lamina ovate or triangular, rarely narrowly
hastate, 5–13·5 × 3·5–11 cm., acute, acuminate, obtuse or emarginate at the apex,
cordate or with rounded basal lobes and a narrow or wide sinus at the base, entire,
membranous, dark glossy green; petiole slender up to 8 cm. long. Flowers in few-
many-flowered cymes, very lax, rarely solitary; peduncle 2–11 cm. long; pedicels and
secondary peduncles 1·5–3 cm. long; bracts minute. Sepals imbricate, ovate-elliptic,
1·3–1·7 cm. long, more or less rounded at the base, acute or rounded at the mucronulate apex, rugulose, with veins showing obliquely in the hyaline margins. Corolla
funnel-shaped, white lilac or pinkish with darker centre, with a more or less cylindrical
tube, 5–8 cm. long, possibly glutinous judging by the appearance of the dried flowers.
Capsule globose, coriaceous, glabrous, tipped by the persistent style. Seeds trigonous,
blackish, glabrous or puberulous.

Zambia. W: Ndola, fl. & fr. 4.ii.1971, *Fanshawe* 11140 *Anton-Smith* s.n. (K; SRGH).
Zimbabwe. N: Gokwe, Malviriviri R., Gokwe-Copper Queen Rd., fl. 5.i.1964, *Bingham* 1050
(K; SRGH). W: Hwange, EU mine, nr. Gwaai R., fl. 27.i.1967, *Wild* 7601 (K; SRGH).
Mozambique. N: Erati, Namapa, nr. Exp. Stat. CICA, fl. & fr. 23.iii.1960, *Lemos & Muácua*
51 (COI; K; LISC; LMA; SRGH). MS: Cheringoma, Inhaminga, Condue-Mazamba,
Nhandinde R., fl. & fr. 30.viii.1966, *Macedo* 2420 (LMA). GI: Gaza, Gondza, fl. 27.ix.1971,
Correia & Marques 2270 (LISC; LMU).

Also in Uganda, Kenya, Tanzania, Zaire and possibly Angola. Open woodland, evergreen
thicket, banks of rivers, roadsides and sandy soils; 150–1036 m.

71. **Ipomoea shirambensis** Baker in Kew Bull. **1894**: 72 (1894).—Dammer in Engl.,
Pflanzenw. Ost-Afr. **C**: 333 (1895).—Hall. f. in Engl., Bot. Jahrb. **28**: 49 (1899)
("*schirambensis*").—Baker & Rendle in F.T.A. **4**, 2: 186 (1905).—Brenan, T.T.C.L.:
20 (1949).—Wild, Guide Fl. Victoria Falls: 156 (1953).—Brenan in Mem. N. Y.
Bot. Gard. **9**: 8 (1954).—Meeuse in Bothalia **6**: 770 (1958).—White, F.F.N.R.: 362
(1962).—Verdc. in F.T.E.A., Convolvulaceae: 134 (1963).—Binns H.C.L.M.: 40
(1968).—Jacobsen in Kirkia **9**: 171 (1973).—Drummond in Kirkia **10**, 1: 271 (1975).
Type: Mozambique, Lower Zambezi, Shiramba, *Kirk* 93 (K, holotype).

Tall perennial woody twiner, up to 7·5 m. high, glabrous or nearly so, often flowering
when completely leafless. Stems glabrous or sometimes pubescent when young, firm
and woody with greyish or yellowish bark and raised longitudinal ridges when old,
oozing sap when cut. Leaves borne on the new shoots, more or less fasciculate,
deciduous; leaf lamina ovate to circular or narrowly triangular, 3·5–8·5 × 1·3–5·5 cm.,
abruptly acute or acuminate and minutely apiculate at the apex, truncate or cordate at
the base, entire, glabrous or more or less pubescent; petiole 3 cm. long. Flowers
appearing before the leaves on the older shoots, few to many, congested in fascicle-like
cymes; peduncle 0–0·5 cm. long; pedicels 10–20 mm. long, thickening upwards from a
slender base; bracts minute. Sepals unequal, oblong-lanceolate, acute and apiculate,
membranaceous, glabrous, the inner ones larger, 9–14 mm. long, corolla funnel-
shaped, white or purple, 3·5–6 cm. long and with the tube constricted at the base,
glabrous. Capsule elongate-ovoid, glabrous, tipped by the persistent style; fruiting
sepals 1·8 cm. long. Seeds elongate-oval, covered with very long spreading brownish
silky hairs up to 1·2 cm. long.

Caprivi Strip. N. of Ngamiland, E. of Kwando R., 945 m., fl. & fr. x.1945, *Curson* 908 (PRE).
Botswana. N: Chobe, between Kasane and Kazungula, fl. & fr. 2.viii.1950, *Robertson & Elffers*
91 (K; PRE; SRGH). SE: nr. Zhilo, Shashi R., more or less c. 8 km. upstream of Shashani R.
junction, fl. & fr. 3.v.1963, *Drummond* 8065 (SRGH). **Zambia**. W: Kitwe, st. 29.viii.1959,
Fanshawe 5141 (K). C: Kabwe, fl. & fr. xi.1928, *van Hoepen* 1421 (PRE). E: nr. foot of Machinje
Hills, 1000 m., fl. & fr. 12.x.1958, *Robson & Angus* 69 (BM; LISC; PRE; SRGH). S:
Livingstone, 884 m., fl. & fr. 9.ix.1955, *Gilges* 424 (PRE; SRGH). **Zimbabwe**. N: Binga,
Sinamwenda Research Stat., Lake Kariba, fl. & fr. 15.x.1971, *Loveridge* 1805 (K; LISC;

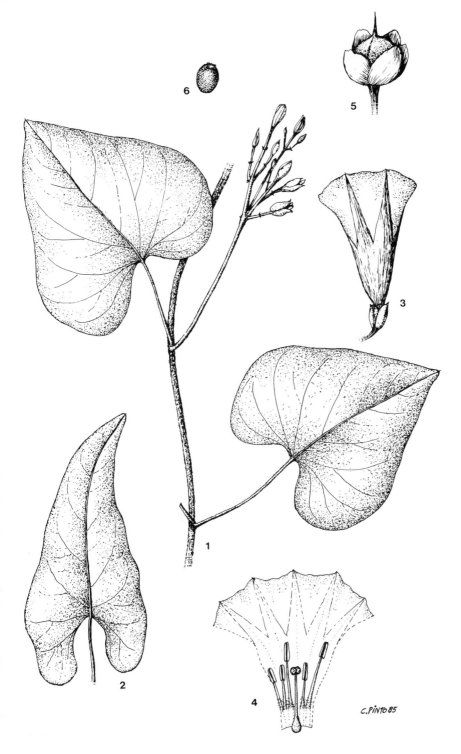

Tab. 29. IPOMOEA SHUPANGENSIS. 1, habit (×½), from *Kirk* 230; 2, leaf (×½); 3, flower (×½); 4, corolla opened to show stamens and pistil (×½), 2–4 from *Lemos & Macuácua* 51; 5, fruit (×½); 6, seed (×1), 5–6 from *Correia & Marques* 2270.

SRGH). W: Hwange, fl. 26.iii.1931, *Eyles* 8053 (BM; K). C: Gweru, c. 6 km. S. of Kwekwe on Kwekwe Reserve Rd., 1280 m., fl. & fr. 8.ix.1965, *Biegel* 178 (SRGH). E: Chimanimani Distr., Changadzi R., Sabi Valley, fl. & fr. 11.ix.1949, *Chase* 1766 (BM; SRGH). S: Buhera, 550 m., fl. & fr. 25.ix.1966, *Plowes* 2791 (K; LISC; SRGH). **Malawi.** N: Nkhata Bay Distr., 8 km. SW. of Nkhata Bay, 580 m., fl. & fr. 11.v.1970, *Brummitt* 10592 (K; LISC). S: Chikwawa, 200 m., fl. 5.x.1946, *Brass* 17992 (BM; K; SRGH). **Mozambique.** N: Mossuril, fl. & fr. 14.viii.1948, *Pedro & Pedrogão* 4787 (LMA). Z: Namagoa, Mocuba, fl. & fr. x.1940, *Faulkner* 79 (BM; COI; K). T: Tete, Songo, 950 m., fl. & fr. 27.x.1973, *Macedo* 5325 (LISC; LMA). MS: Gorongosa, Nat. Park, fl. & fr. 2.ix.1965, *Macedo* 1244 (LMA). GI: Caniçado, 20 km. from Lagoa Nove to Aldeia da Barragem, fl. & fr. 17.vii.1969, *Correia & Marques* 922 (LISC; LMU). M: Sábiè, 30 km. from Chinhanguanine to Moamba, fl. & fr. 22.vi.1970, *Correia & Marques* 1825 (LISC; LMU).

Also from Tanzania and S. Africa (Transvaal). Woodland, deciduous thicket, bushland, roadside and sandy soils; 30–1400 m.

72. **Ipomoea rubens** Choisy in Mem. Soc. Phys. Genève **6**: 436 (1834).—Prain in Journ. Asiatic Soc. Bengal **63**: 109 (1894).—Chiov., Miss. Stef.—Paoli, Collez. Bot. **1**: 125 (1916).—Verdc. in F.T.E.A., Convolvulaceae: 134 (1963).—Roessler in Merxm., Prodr. Fl. SW. Afr. **116**: 16 (1967).—Binns, H.C.L.M.: 40 (1968).—Gibbs-Russell in Kirkia **10**: 490 (1977). Type from India.

    Ipomoea lilacina Bl., Bijdr. Fl. Ned. Ind. **13**: 716 (1826).—Hall. f. in Bull. Soc. Bot. Belg. **37**: 100 (13 Aug 1898).—Baker & Rendle in F.T.A. **4**, 2: 187 (1905).—N.E. Br. in Kew Bull. **1909**: 122 (1909).—Eyles in Trans. Roy. Soc. S. Afr. **5**: 455 (1916).—Wild, Guide Fl. Victoria Falls: 156 (1953).—Dandy in F. W. Andr. Fl. Pl. Anglo-Egypt. Sudan **3**: 119 (1956) *nom. illegit.* non Schrank (1822). Type from Java.

    Ipomoea riparia G. Don, Gen. Syst. **4**: 265 (1837).—Exell, Cat. Vasc. Pl. S. Tome: 251 (1944).—Brenan, T.T.C.L.: 172 (1949).—van Ooststr. in Fl. Males. Ser. 1, **4**: 484 (1953).—Brenan in Mem. N.Y. Bot. Gard. **9**: 8 (1954).—Meeuse in Bothalia **6**: 766 (1958).—White, F.F.N.R.: 362 (1962).—Heine in F.W.T.A. ed. 2, **2**: (1963). Type from S. Tomé.

    Ipomoea baclei Choisy in Mem. Soc. Phys. Geneve **8**: 60, t. 2 (1838), as "*baclii*". Type from Senegal.

    Pharbitis fragans Choisy in DC., Prodr. **9**: 341 (1845). Type from Madagascar.

    Ipomoea lindleyi Choisy in DC., Prodr. **9**: 371 (1845).—Schinz in Pl. Menyhart: 69 (1905).—Gomes e Sousa in Bol. Soc. Moçambique **32**: 86 (1936). Type from Madagascar.

    Ipomoea fragans (Choisy) Hall. f. in Engl., Bot. Jahrb. **18**: 153 (1893).—Dammer in Engl., Pflanzenw. Ost–Afr. C: 333 (1895).—van Ooststr. in Blumea **3**: 564 (1940).

    Ipomoea oxyphylla Baker in Kew Bull. **1894**: 71 (1894). Type from Angola.

    Ipomoea stuhlmannii Dammer in Engl., Pflanzenw. Ost–Afr. **C**: 333 (1895). Type from Tanzania.

    Ipomoea brasseuriana De Wild. in Ann. Mus. Congo, Sér. 4: 115 (1903).—Baker & Rendle in F.T.A. **4**; 2: 186 (1905). Fries, Wiss. Ergebn. Schwed. Rhod.-Kongo-Exped. **1**: 270 (1916). Type from Zaire.

Perennial twiner, up to 4 m. long. Stems terete, rather woody, finely striate when dry densely short-pilose with soft white hairs. Leaf lamina broadly ovate to circular, 5–15 × 4–12 cm., acuminate at the apex, cordate at the base, entire or rarely 3-lobed, glabrous to slightly pubescent or densely velutinous; petiole slender, 3–7 cm. long, pilose like the stems. Inflorescences 1–few-flowered cymes with very short cymes branches and consequently flowers subumbellate; peduncle 2–15 cm. long, pilose like the stems; pedicels up to 1·5 cm. long, pilose, bracts ovate, minute. Sepals ovate, obtuse or minutely mucronate, 6–8 mm. long, pilose. Corolla funnel-shaped, purple or mauve with darker centre, 4–5 cm. long, sparsely pilose. Capsule globose, glabrous. Seeds ovoid, densely hairy, hairs 2·5 mm. long.

Botswana. N: N. of Okavango Swamps, between Moanachira and Dobe Lagoons, 950 m., st. 8.iii.1972, *Biegel & Gibbs-Russell* 3916 (K; SRGH). **Zambia.** B: Siwelewele, Mashi R., 1036 m., fl. 8.viii.1952, *Codd* 7429 (BM; K; PRE). N: Chiengi Distr., Luchinda R. Bridge to Pweto Rd., 950 m., fl. & fr. 25.v.1961, *Astle* 674 (K; SRGH). S: Victoria Falls, Long Island 8 km. above Falls, 914 m., fl. & fr. v. 1904, *Eyles* 140 (BM; SRGH). **Zimbabwe.** N: Binga, Milibizi R., fl. & fr. 22.iii.1971, *Tur* 1423 (SRGH). **Malawi.** N: Mzimba Distr., Njakwa, east end of S. Rukuku Gorge, 6 km. E. of Rumphi, 1040 m., fl. 6.vii.1970, *Brummitt* 11842 (K; LISC; MAL). S: Nsanje Distr., Chiromo Ferry, 28.v.1970, *Brummitt* 11127 (K; LISC; MAL; PRE; SRGH).). **Mozambique.** Z: Maganja da Costa, Errive R., 13 km. of Maganja da Costa, fl. & fr. 23.vi.1946, *Pedro* 1443 (LMA). T: Tete, Msusa, 213 m., 25.viii.1950, *Chase* 2841 (BM; K; SRGH).

Also in S. Tomé, Uganda, Kenya, Tanzania, Angola, Namibia; widespread in tropical Africa, Madagascar, Mascarene Isl. and also in India, Malaysia, Philippines and Guiana. Swamps by rivers and marshy bushlands; 5–1220 m.

73. **Ipomoea mauritiana** Jacq., Collect. **4**: 216 (1791); in Hort. Schoenbr. **2**: 39, t. 200 (1800).—Heine in F.W.T.A. ed. 2, **2**: 351, fig. 284 (1963).—Verdc. in F.T.E.A., Convolvulaceae: 135 (1963).—Munday & Forbes in Journ. S. Afr. Bot. **45**: 9 (1979). Type a plant from Mauritius cultivated at Vienna, probably not preserved.

Convolvulus paniculatus L., Sp. Pl.: 156 (1753). Type from India.

Ipomoea paniculata (L.) R. Br., Prodr.: 486 (1810).—Hall. f. in Engl., Bot. Jahrb. **18**: 149 (1893) *nom. illegit.* non Burm. (1768).

Ipomoea eriosperma Beauv., Fl. Oware **2**: 73, t. 105 (1819). Type from Ghana.

Ipomoea digitata sensu Baker & Rendle in F.T.A. **4**, 2: 189 (1905).—Hutch. & Dalz., F.W.T.A. **2**: 216, fig. 251 (1931).—van Ooststr. in Fl. Males., Ser. 1, **4**: 483, fig. 55 (1953).—Dandy in F. W. Andr., Fl. Pl. Anglo-Egypt. Sudan **3**: 119 (1956).—Meeuse in Bothalia **6**: 767 (1958).—Ross. Fl. Natal: 296 (1972) non L. (1759).

Ipomoea digitata sensu Baker & Rendle var. *eriosperma* (Beauv.) Rendle in F.T.A. **4**, 2: 190 (1905) (Sphalm. "var. *eriocarpa*").—van Ooststr. in Fl. Males., Ser. 1, **4**: 484 (1953).

Large glabrous perennial twiner, occasionally prostrate, with large tuberous roots. Stems twining, terete, becoming woody. Leaf lamina circular in outline, entire or palmately 3–9-lobed, 5–20 × 6–15 cm., cordate or truncate at the base; lobes lanceolate to ovate, acuminate, minutely mucronate, at apex, entire; petiole 3–11 cm. long, smooth or muriculate. Inflorescence few to many-flowered; peduncle 2·5–18 cm. long, terete but often angular and cymosely branched near the apex; pedicels terete, 9–25 mm. long; flower-buds globular. Sepals markedly convex, clasping the corolla tube, orbicular or elliptic, 6–12 mm. long. Corolla funnel-shaped with spreading limb, and narrow tube below, reddish-purple or mauve with darker centre, 5–6 cm. long. Capsule ovoid or globose, obtuse, glabrous. Seeds black, covered with silky hairs more or less 7 mm. long.

Zambia. N: from Kawambwa to Mbereshi Rd., fl. & fr. 13.iii.1969, *Anton-Smith* s.n. (SRGH). W: Kitwe, fr. 12.iii.1968, *Fanshawe* 10276 (K; SRGH). E: Nyika Plateau, fl. & fr. ii–iii.1903, *McClounie* 143 (K). S: Namwala, Msusa-Kafue Confluence, Kafue National Park, fl. & fr. 12.i.1964, *Mitchell* 24/55 (K; LISC; SRGH). **Zimbabwe.** N: Chirundu, 11 km. downstream, fl. & fr. 20.I.1967, *Wild* 7594 (K; LISC; SRGH). W: Victoria Falls, 914 m., fl. & fr. 31.i.1934, *Saunders Davies* s.n. (BM). **Malawi.** N: Nkhata Bay Distr., Sanga, c. 16 km. S. of junction with Mzuzu-Nkhata Bay Rd., 488 m., fl. & fr. 17.xii.1972, *Pawek* 6112 (K). **Mozambique.** Z: Morrumbala, 19 km. from Morrumbala to Mopeia, fl. & fr. 24.ii.1972, *Correia & Marques* 2778 (K; LISU; LMU; SRGH). T: Tete, Cahobra Bassa, Morumbuè Mt., 950 m., fl. & fr. 20.xii.1973, *Macedo* 5470 (LISC; LMU). GI: Zavala, Quissico, 28.ii.1955, *Exell, Mendonça & Wild* 704 (BM; LISC; SRGH). M: Maputo, 7 km. from Ponta do Ouro to Zitundo, fr. 9.vii.1971, *Correia & Marques* 2108 (COI; LISC; LMA; LMU; SRGH).

West and East tropical Africa, circumtropical; also in S. Africa (Natal). Open and riparian woodland, savanna with trees, margin and banks of rivers, swamps, sandy soils and roadside; 20–2440 m.

74. **Ipomoea violacea** L., Sp. Pl.: 161 (1753).—Manitz in Fedde, Repert. **88**: 265: 271 (1977).—Verdc. in Kew Bull. **33**: 167 (1978). Type from South America: illustration of Plumier Plantarum Americanum t. 3 (1703), (P, lectotype).

Convolvulus grandiflorus Jacq., Hort. Vindob. **3**: 39, t. 69 (1776).—Desr. in Lam., Encycl. **3**: 543 (1791). Type a specimen cultivated at Vienna from seed from St. Peter's River, Martinique, probably not preserved (W?).

Ipomoea macrantha Roem. & Schultes, Syst. Veg. **4**: 251 (1819).—Gunn in Brittonia **24**: 158, fig. 3 (1972). Type from Australia.

Convolvulus tuba Schlechtend. in Linnaea **6**: 735 (1831). Type from the West Indies.

Calonyction grandiflorum (Jacq.) Choisy in Mem. Soc. Phys. Genève **6**: 442 (1834).

Ipomoea glaberrima Hook. in Hook., Journ. Bot. **1**: 357 (1834). Type from Seychelles, Bojer (K, lectotype).

Ipomoea tuba (Schlechtend.) G. Don, Gen. Syst. **4**: 271 (1837).—van Ooststr. in Blumea **3**: 575 (1940) & in Fl. Males., Ser. 1, **4**: 487 (1953).—Verdc. in Kew Bull. **13**: 216 (1958) & in F.T.E.A., Convolvulaceae: 137 (1963).

Ipomoea grandiflora (Jacq.) Hall. f. in Engl., Bot. Jahrb. **18**: 153 (1893) *nom. illegit.* —Baker & Rendle in F.T.A. **4**, 2: 190 (1905) non (L. f.) Lam. (1797).

A glabrous perennial herb. Stems twining or prostrate, terete or angular, often longitudinally wrinkled but otherwise smooth, ochraceous. Leaf lamina circular to ovate, 5–16 × 5–14 cm., acuminate or cuspidate and mucronulate at the apex, deeply cordate at the base; petiole 3·5–8 cm. long. Flowers usually solitary, rarely in few-flowered cymes; peduncle 0·75–7 cm. long; pedicels 1·5–3 cm. long, becoming much thickened in fruit; bracts small, deciduous. Sepals subequal, elliptic, 1·6–2 cm.

long, attaining 3 cm. in fruit, obtuse at the apex, margins hyaline, coriaceous. Corolla salver-shaped or very narrowly funnel-shaped, white and/or pale greenish-yellow, opening about mid-night; tube 7–8·5 cm. long; limb 2–4 × 5–6 cm., Capsule globose, 2–3 cm. in diam. Seeds subtrigonous more or less 1 cm. long, black, densely short tomentose and with a narrow ridge of longer hairs more or less 3–6 mm. long.

Mozambique. N: Palma, Rovuma Bay, fl. & fr. iii. 1861, *Kirk* s.n. (K).
Also in Ghana, Kenya, Tanzania, Mascarene Isl., Seychelles; tropical Asia to Polynesia and also West Indies and Florida. Coastal bushland.

75. **Ipomoea albivenia** (Lindl.) Sweet, Hort. Brit., ed. 2: 372 (1830).—Hall f. in Engl., Bot. Jahrb. **28**: 151 (1899).—Macnae & Kalk, Nat. Hist. Inhaca I., Moçamb.: 152 (1958).
—Meeuse in Bothalia **6**: 768 (1958).—Verdc. in F.T.E.A., Convolvulaceae: 140 (1963).
—Ross, Fl. Natal: 295 (1972).—Drummond in Kirkia **10**, 1: 271 (1975).—Compton, Fl. Swaziland: 476 (1976).—Munday & Forbes in Journ. S. Afr. Bot. **45**: 9 (1979). Type a cultivated plant grown from seed from Mozambique, Delagoa Bay, coll. *Forbes*, probably not preserved.
   Convolvulus albivenius Lindl. in Bot. Reg. 13, t. 1116 (1827).
   Ipomoea gerrardii Hook. f. in Bot. Mag. **93**: t. 5651 (1867).—Hall. f. in Engl., Bot. Jahrb. **18**: 151 (1893). & **28**: 51 (1899). Type a cultivated plant grown at Kew from seed from Natal, coll. *Sutherland*, probably not preserved.
   Ipomoea wakefieldii Baker in Kew Bull. **1894**: 73 (1894).— Dammer in Engl., Pflanzenw. Ost-Afr. **C**: 333 (1895).—Baker & Rendle in F.T.A. **4**, 2: 184 (1905). Type from Kenya.

Climbing shrub up to 10 m. long, close to the following species but distinct because the different indumentum and distribution. Stems slender, trailing when young herbaceous with white somewhat floccose tomentum, soon glabrous, older ones becoming woody. Leaf lamina circular ovate to oblong-ovate, up to 13 × 16 cm., cuspidate or emarginate at the apex, more or less cordate at the base, entire to sinuous or somewhat crenate, when very young covered on both sides with a white floccose tomentum, reticulate beneath due to a dense tomentum on the main nerves and veins, glabrescent, petiole slender, 3·5–7·5 cm. long, white tomentose. Flowers few in a lax cyme; peduncle short, tomentose; pedicels short, tomentose; bracts obovate, 1·6 cm. long, floccosely tomentose outside, deciduous. Sepals subequal, broadly oblong, 1·5 cm. long, obtuse, chartaceous, thinly covered with scattered tomentum, glabrescent, generally retaining the pubescence longest at the base, in fruit ultimately spreading to reflexed. Corolla funnel-shaped, White, cream or yellowish, with mauve tube up to 14·5 cm. long, glabrous. Capsule 1·6–2·2 cm. long, apiculate, glabrous. Seeds brown densely covered with very long whitish cottony hairs, giving the dehisced capsule the appearance of a ripe cotton ball.

Zimbabwe. S: Beit Bridge, Chiturupadzi Dip Camp Area, c. 88 km. E. of Beit Bridge, fr. 18.iii.1967, *Mavi* 239 (K; LISC; SRGH). **Mozambique**. Z: Maganja da Costa, 305 m., 1908, *Sim* 21091 (PRE). MS: Chimoio, Tembe, Chindaza Mt., fl. & fr. 16.iii.1948, *Garcia* 623 (C; LISC; LMA; WAG). GI: Chibuto, from Maniquenique to Licilo, nr. Licilo, fl. & fr. 13.ii.1959, *Barbosa & Lemos* 8396 (COI; K; LISC; LMA). M: Maputo, Costa do Sol, fl. & fr. 2.iii.1960, *Balsinhas* 123 (COI; K; LISC; LMA).
Also in Kenya, S. Africa (Transvaal, Swaziland and Natal). Woodland, savana with trees, moist ground, coastal bushland and sandy soil: 40–800 m.

76. **Ipomoea verbascoidea** Choisy in Mem. Soc. Phys. Genève **8**: 56 (1838) & in DC., Prodr. **9**: 356 (1845).—Hall. f. in Engl., Bot. Jahrb. **18**: 151 (1893).—Hiern, Cat. Afr. Pl. Welw. **1**: 741 (1898).—Hall. f. in Bull. Soc. Roy. Bot. Belg. **37**: 100 (1898).—Baker & Rendle in F.T.A. **4**, 2: 183 (1905).—Eyles in Trans. Roy. Soc. S. Afr. **5**: 456 (1916).—Dandy in F. W. Andr., Fl. Pl. Anglo-Egypt. Sudan **3**: 120 (1956).—Meeuse in Bothalia **6**: 769 (1958).—Verdc. in Kew Bull. **13**: 215 (1958).—White, F.F.N.R.: 362 (1962).—Verdc. in F.T.E.A., Convolvulaceae: 140, fig. 22, 11 and 24, 8 (1963). Heine in F.W.T.A. ed. 2, **2**: 1348 (1963).—Roessler in Merxm. Prodr. Fl. SW. Afr. **116**: 17 (1967).—Verdc. in Kirkia **6**: 121 (1967).—Jacobsen in Kirkia **9**: 171 (1973).—Drummond in Kirkia **10**, 1: 271 (1975). Type from Angola.
   Ipomoea elliottii Baker in Kew Bull. **1894**: 69 (1894). Type: Zimbabwe, "Matabeleland", W. *Elliott* s.n. (K, holotype).
   Ipomoea dammarana Rendle in Journ. Bot. **34**: 36 (1896).—Baker & Rendle in F.T.A. **4**, 2: 183 (1905).—Eyles in Trans. Roy. Soc. S. Afr. **5**: 454 (1916).—Suessenguth & Merxm. in Trans. Rhod. Sci. Ass. **43**: 38 (1951). Type from Namibia.

Ipomoea lukafuensis De Wild. in Ann. Mus. Congo Belge, Bot. Sér. **4**: 112 (1903). Type from Congo.
   Ipomoea couceiroi Rendle in Journ. Bot. **46**: 182 (1908). (*"conceiroi"*). Type from Angola.
   Ipomoea assumptae Mattei in Malpighia **31**: 147 (1928). Type a specimen grown in Italy from a seed collected in Bulawayo (Zimbabwe).
   Acmostemon angolensis Pilger in Notizbl. Bot. Gard. Berl. **13**: 106 (1936). Type from Angola.

Suberect to climbing shrub. Stems erect, decumbent or rambling, up to 1·5 m. tall covered with a yellowish or white woody somewhat floccose tomentum as are the leaves, petioles, peduncles, bracts and generally the sepals. Leaf lamina oblong or ovate-circular 5–18·5 × 5–17·5 cm. obtuse or subacute, shortly apiculate, rarely acuminate at the apex, cordate or truncate at the base, entire or slightly sinuate, glabrous or with floccose tomentum above and covered with a dense granular tomentum beneath; petiole 2–10 cm. long, bearing a gland on each side of the insertion of the lamina. Flowers solitary or in more or less 3-flowered cymes; peduncle 1–2·5 cm. long; pedicels about 1·2 cm. long; bracts linear-oblong to linear-oblanceolate, 1·3–2 cm. long, tomentose outside. Sepals elliptic, 1–2 cm. long, obtuse, chartaceous, generally persistently tomentose. Corolla funnel-shaped, white, rose-purple or more or less mauve with deeper mauve throat, 6·5–11 cm. long, tube broad cylindrical up to 6 cm. long and limb 5–7 cm. in diam. Capsule oblong ovoid or globose, 20–25 mm. long, coriaceous glabrous. Seeds ovoid, brown, densely covered with long white or sometimes fulvous cottony hairs, giving the dehisced fruit the appearance of an open cotton ball.

Caprivi Strip. Ngoma Area, 930 m., fl. 20.xii.1958, *Killick & Leistner* 3008 (SRGH). **Botswana.** N: Dobe, fl. & fr. 28.xii.1964, *Lee* 104 (SRGH). **Zambia.** B: Sesheke, N. of Machili, fl. 19.xii.1952, *Angus* 954 (FHO; K). N: Kasama Distr., Mpika Rd. nr. Chibutubutu, 1320 m., fl. 27.ii.1960, *Richards* 12635 (K; SRGH). W: c. 96 km. S. of Mwinilunga on the Kabompo Rd., fr. 3.vi.1963, *Loveridge* 731 (K; LISC). C: Kabwe, fl. & fr. 16.xii.1957, *Fanshawe* 4134 (FHO; K). S: Livingstone Distr., Katambora, fl. 29.ii.1956, *Gilges* 683 (SRGH). **Zimbabwe.** N: Gokwe Distr., nr. Gokwe, fl. 27.xii.1962, *Bingham* 361 (K; SRGH). W: Bulawayo, 1·5 km. S. of Old Essexvale Rd., fr. 1.ii.1975, *Cross* 264 (K; SRGH). C: Marondera Distr., c. 30 km. S. of Marondera on Farm Monte Cristo, c. 1200 m., fl. 2.ii.1973, *Biegel* 4137 (K; LISC; SRGH). E: Mutare, Dora R. at bridge, 1036 m., fl. 20.ii.1958, Chase 6829 (COI; K; SRGH). S: Masvingo Distr., Zimbabwe Great National Park, fl. 3.ii.1971, *Chiparawasha* 339 (SRGH). **Malawi.** N: Mzimba Distr., c. 9·6 km. S. of Njakwa, 7 mls N. of Mzuzu junction, 1158 m., fl. 2.iv.1977, *Pawek* 12544 (SRGH). C: Kasungu, Chipala Hill, c. 6 km. N. of Kasungu, 1000 m., fl. 14.i.1959, *Robson* 1180A (BM; K; LISC). S: Kasupe Distr., Chikala Hills, 1065 m., fl. & fr. 17.ii.1975, *Brummitt & al.* 14364 (K; SRGH). **Mozambique.** T: Tete, Songo, fr. 18.iv.1972, *Macedo* 5208 (LISC; LMA; LMU). MS: Chimoio, Belas Mts., fr. 2.iv.1948, *Garcia* 839 (LISC; LMA).
   Also in Sudan, Uganda, Tanzania, Zaire, Angola and Namibia. Woodland, bushland, open grassland and Kalahari sands; 600–1320 m.

77. **Ipomoea bakeri** Britten in Journ. Bot. **1894**: 85 (1894).—Dammer in Engl., Pflanzenw. Ost-Afr. **C**: 333 (1895).—Baker & Rendle in F.T.A. **4**, 2: 184 (1905).—Verdc. in Kew Bull. **14**: 341 (1960).—White, F.F.N.R.: 361 (1962). Type: Zambia, Lake Tanganyika, *Carson* 18 (K, holotype).
   Ipomoea discolor Baker in Kew Bull. **1894**: 69 (1894) *nom. illegit.* non G. Don (1837).

Arching shrub, up to 2·5 m. high, close to the species above, but distinct because of the different shape of the leaves and distributions. Stems slender, trailing, woody, covered with whitish woolly somewhat floccose tomentum as are the leaves, petioles, peduncles, bracts and generally the sepals. Leaf lamina ovate-oblong, elliptic-oblong or lanceolate-oblong, 5–20·5 × 3–13 cm., obtuse or acute at the apex broadly rounded or subtruncate at the base, margin entire, repand to profoundly dentate, glabrescent above with an impressed nervation and covered with a dense granular whitish tomentum beneath; petiole up to 4 cm. long, bearing a gland on each side of the insertion of the lamina. Flowers solitary or in more or less 3-flowered cymes; peduncle 1–3 cm. long; pedicels about 1·2 cm. long; bracts obovate, up to 2·5 cm. long, tomentose outside. Sepals elliptic, up to 1·6 cm. long, obtuse, chartaceous, generally persistently tomentose. Corolla funnel-shaped, white or pale mauve, with purple throat, 6·5–9 cm. long and limb up to 6 cm. in diam. Capsule broadly ellipsoid, up to 1·6 cm. long, coriaceous, glabrous. Seeds densely covered with long white or sometimes fulvous cottony hairs, giving the dehisced fruit the appearance of an open cotton ball.

Zambia. N: Misamfu, 6 km. N. of Kasama, fr. 4.iv.1961, *Angus* 2668 (FHO; K; SRGH). Not known elsewhere. Woodland, bushland, grassland, dry ground and sandy soil; 1320–1680 m.

78. **Ipomoea prismatosyphon** Welw. in Ann. Concelho Ultram. 1858 [Apont. Phyto-Geogr.]: 585 (1859).—Britten in Journ. Bot. **32**: 84 (1894).—Rendle in Journ. Bot. **32**: 215 (1894) (*"prismatosiphon"*).—Baker & Rendle in F.T.A. **4**, 2: 181 (1905).—Verdc. in F.T.E.A., Convolvulaceae: 142 (1963). TAB. 30. Type from Angola.

Erect shrubby herb, up to 2 m. tall, lactiferons. Stems many, stout, unbranched, ridged, angular, woody, tomentose or glabrous. Leaf lamina oblong to ovate, 7–14·5 × 3–11·5 cm., rounded or acute and shortly apiculate at the apex, rounded or cuneate at the base, with deep venation, entire, undulate, sometimes deeply dentate, or deeply trilobed in var. *trifida* Verdc., upper surface puberulous or tomentose, lower one covered with a dense floccose tomentum or entirely glabrous all over; petiole 1–5 cm. long, biglandular at the apex. Flowers solitary, showy and scented; peduncle 1·2–4 cm. long, more or less pubescent; pedicel up to 1 cm. long; bracts obovate, foliaceous, 1·6–4 × 1·3–3 cm., tomentose or glabrous, venose, hiding the calyx in the buds, but soon falling. Sepals elliptic, obtuse, tomentose or glabrous, 1·7–3·2 cm. long. Corolla funnel-shaped, white, pale rose or purplish at the base of the tube, 9–16 cm. long; tube broad cylindrical up to 2 cm. in diam. and 8 cm. long; limb glabrous save for sparsely lanate apical part of the midpetaline areas outside. Capsule ovoid, 3 cm. long, woody. Seeds ovoid, 9 mm. long covered with long shiny pale tawny hairs.

Var. **prismatosyphon**

Ipomoea buchneri Peter in Engl. & Prantl., Pflanzenfam. **4**, 3a: 29 (1891).—Hall. f. in Engl., Bot. Jahrb. **18**: 151 (1893). Type from Angola.
Ipomoea buchneri Peter var. *tomentosa* Hall. f. in Engl., Bot. Jahrb. **18**: 152 (1893). Type as for *Ipomoea prismatosyphon* Welw.
Ipomoea magnifica Hall. f. in Engl. Bot. Jahrb. **18**: 152 (1893).—Baker & Rendle in F.T.A. **4**, 2: 181 (1905).—Dandy in F. W. Andr. Fl. Pl. Anglo-Egypt. Sudan **3**: 119 (1956). Type from Sudan.
Ipomoea prismatosyphon Welw. var. *buchneri* (Peter) Britten [sphalm *buchingeri*] in Journ. Bot. **32**: 85 (1894).—Baker & Rendle in F.T.A. **4**, 2: 181 (1905).
Ipomoea buchneri Peter var. *latifolia* Hall. f. in Bull. Soc. Roy. Bot. Belg. **37**: 100 (1898). Type as for *Ipomoea magnifica* Hall. f.
Ipomoea hanningtonni (Baker) Rendle in F.T.A. **4**, 2: 182 (1905).—Brenan in Kew Bull. **1949**: 93 (1949).—Verdc. in Kew Bull. **13**: 215 (1958) pro parte, *nomen confusum*, non Baker (1894).
Ipomoea tessmannii Pilger in Notizbl. Bot. Gart. Berl. **7**: 542 (1921).—Hutch. & Dalz., F.W.T.A. **2**: 215 (1931). Type from Cameroon.

Zambia. N: Mbala, Mpulungu, Lake Tanganyika, fr. 19.vii.1930, *Pole Evans* 2995 (K; PRE). **Malawi.** N: Chitipa, between Wenya and Chitsanga, fl. & fr. 14.iii.1977, *Grosvenor & Renz* 1223 (K; SRGH).
Also in Sudan, Nigeria, Zaire and Cameroun, Uganda, Tanzania and Angola. Woodland, bushland, grassland and roadsides; 1500–1615 m.
The var. *trifida* has upper leaves deeply trilobed. It occurs in Tanzania.

79. **Ipomoea adenioides** Schinz in Verhandl. Bot. Ver. Brandenb. **30**: 270 (27th Sept. 1888).—Baker & Wright, F.C. **4**, 2: 51 (1904).—Baker & Rendle in F.T.A. **4**, 2: 195 (1906).—N.E. Br. in Kew Bull. **1909**: 122 (1909).—Meeuse in Bothalia **6**: 770 (1958). —Verdc. in Kew Bull. **14**: 342 (1960)—Roessler in Merxm. Prodr. Fl. SW. Afr. **116**: 12 (1967). TAB. **31**. Type from Namibia.

Erect shrub, up to 1·20 m. high. Stems covered with short white silky hairs when young, older ones glabrous, greyish or yellowish canescent. Leaves at the ends of the young branches; leaf lamina from lanceolate or oblanceolate to elliptic, obovate to obovate-circular (ovate-lanceolate in var. *ovato-lanceolata*) obtuse to acute often with a white-hairer mucro, usually cuneate at the base, 3–8 × 2–5·5 cm., entire, green, glabrescent above, densely white-silky beneath mainly on the veins, ultimately glabrescent; petiole up to 3 cm. long, shortly silky-pubescent. Flowers solitary, peduncle up to 1 cm. long, densely silky-pubescent as the pedicels, bracts and calyx; pedicels up to 0·5 cm. long; bracteoles linear to linear-lanceolate, acute, up to 1·8 cm. long. Sepals subequal, lanceolate, very long acuminate, up to 2·8 cm. long in fruit. Corolla very long salver-shaped, white or pink, deep magenta inside, up to 11 cm. long, appressed, silky outside; tube narrowly cylindric 7–9 cm. long; limb spreading, up to 6

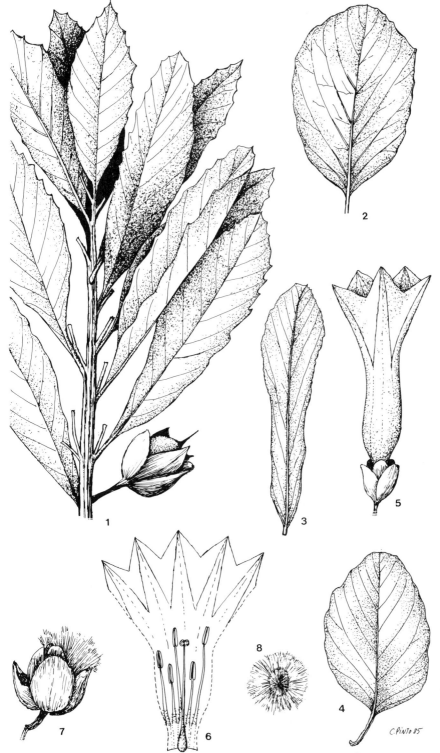

Tab. 30. IPOMOEA PRISMATOSYPHON. 1, habit ($\times\frac{1}{2}$), from *Drummond & Williamson* 9969; 2–4, leaves ($\times\frac{1}{2}$), 2 & 4 from *Brummitt & Banda* 16213; 3, from *Nash* 152; 5, flower ($\times\frac{1}{2}$); 6, corolla opened to show stamens and pistil ($\times\frac{1}{2}$), 5–6 from *Nash* 152; 7, fruit, ($\times\frac{1}{2}$); 8, seed ($\times\frac{1}{2}$), 7–8 from *Pole-Evans* 2966.

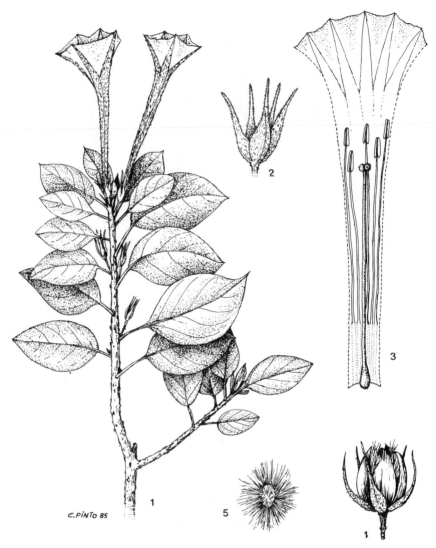

Tab. 31. IPOMOEA ADENIOIDES. 1, habit (×½); 2, calyx (×1); 3, corolla opened to show stamens and pistil (×1), 1–3 from *Lugard* 224; 4, fruit (×1); 5, seed (×1), 4–5 from *Allen* 76.

cm. in diam. Capsule ovoid, up to 2·3 cm. long, thinly hairy to glabrous, coriaceous. Seeds densely covered with very long shiny fulvous-cream hairs.

Var. **adenioides**

Ipomoea marlothii Engl., Bot Jahrb. **10**: 244 (9th Oct. 1888). Type from Namibia.
Rivea adenioides (Schinz) Hall. f. in Engl., Bot. Jahrb. **18**: 156 (1893).

Botswana. N: Quangwa, 27 km. NE. of Aha Hills, fr. 12.iii.1965, *Wild & Drummond* 6933 (K; LISC; SRGH). SW: Kgalagadi, Mabua, Sefhubi pan 1·006 m., fl. 28.ii.1963, *Leistner* 3100 (K; PRE). SE: Orapa, fr. 30.iii.1974, *Allen* 76 (K; PRE). **Zimbabwe**. N: Gokwe, fr. v. 1960, *Thompson* 26/60 (SRGH). W: Hwange, 732 m., fl. & fr. 6.i.1935, *Eyles* 8435 (K; SRGH).
Also in Somalia, Angola, Namibia and S. Africa (Transvaal). Limestone pavement and sandy loam soil; 900–1010 m.
The var. *ovato-lanceolata* Hall. f. has leaves ovate-lanceolate. It occurs in Ethiopia and Somalia.

80. **Ipomoea kituiensis** Vatke in Linnaea **43**: 511 (1882).—Baker & Rendle in F.T.A. **4**, 2: 196 (1906).—Brenan, T.T.C.L.: 11 (1949).—Verdc. in Kew Bull. **13**: 216 (1958).—White, F.F.N.R.: 362 (1962).—Verdc. in F.T.E.A., Convolvulaceae: 146, fig. 25 (1963).— Binns, H.C.L.M.: 40 (1968).—Drummond in Kirkia **10**, 1: 271 (1975). Type from Kenya.

Shrub probably with tuberous roots. Stems suberect or twining, up to 6 m. long, pubescent, with leaf scars prominent. Leaf lamina ovate to reniform, 3·5–17·5 × 3·2–17·5 cm., obtuse, apiculate, acuminate or even bilobed at the apex, cordate at the base, sparsely pubescent or glabrous above, finely pubescent or glabrescent beneath; petiole 1–17·5 cm. long, pubescent. Flowers few to many more or less crowded in cymes; peduncle 3·5–13 cm. long, pubescent; pedicels and secondary peduncles 0·2– 2 cm. long, pubescent; bracts linear-lanceolate 1–1·5 cm. long, white, silky appressed pubescent as in the sepals. Sepals linear-lanceolate to ovate, long, tapering upwards. Corolla funnel-shaped, white, cream or yellow with a purple centre, 5–8 cm. long; midpetaline areas pubescent or pilose; tube narrowed below. Capsule ellipsoid, 1·7–2 cm. long. Seeds ovoid, 9 mm. long covered with long golden hairs about 10 mm. long.

Var. **kituiensis**
   Rivea kituiensis (Vatke) Hall. f. in Engl., Bot. Jahrb. **18**: 156 (1893).
   Ipomoea tambelensis Baker in Kew Bull. **1894**: 72 (1894).—Dammer in Engl., Pflanzenw. Ost-Afr. **C**: 333 (1895). Type: Malawi, Upper Shire Valley, Tambala, *Kirk* s.n. (K, holotype).
   Ipomoea nyikensis Hall. f. in Engl., Bot. Jahrb. **28**: 53 (1899).—Baker & Rendle in F.T.A. **4**, 2: 196 (1906). Type from Tanzania.

   Zambia. N: Luangwa Valley, nr. Mfuwe, 690 m., fl. & fr. 16.iii.1966, *Astle* 4644 (K; SRGH). W: Kitwe, st. 7.vi.1963, *Fanshawe* 7917 (K). S: Bombwe Forest, Livingstone-Machili, fl. & fr. 25.iii.1933, *Martin* 644 (FHO; K). **Zimbabwe**. N: Gokwe, nr. Iare R., fr. 19.iv.1963, *Bingham* 633 (K; SRGH). W: Hwange c. 19 km. NE. of Kamativi, 914 m., fl. & fr. 8.iii.1964, *Leach* 12130 (COI; K; LISC; SRGH). C: Harare, 1463 m., fl. & fr. 8.iv.1926, *Eyles* 1210 (K). E: Mutare, Mpembi Mt., 914 m., st. 16.iii.1964, *Chase* 8134 (COI; K; LISC). S: Bikita, Devuli Ranch, c. 16 km. N. of Umkondo Mine, fr. 21.v.1972, *Leach & Cannell* 14893 (K; LISC; SRGH). **Malawi**. S: nr. Tambala, Upper Shire Valley, fl. iv.1859, *Kirk* s.n. (K).
   Also in Ethiopia, Kenya and Tanzania. Woodland, savanna and sandy soil; 610–1140 m.
   The var. *massaiensis* (Pilger) Verdc. has sepals broadly triangular to ovate, 7–10 × (3·5) 4·5– 6 mm. It occurs in Tanzania.

81. **Ipomoea consimilis** Schulze-Menze in Notizbl. Bot. Gart. Berl. **14**: 112 (1938).—Brenan, T.T.C.L.: 170 (1949).—Verdc. in F.T.E.A., Convolvulaceae: 149 (1963). Type from Tanzania.

Climbing subshrub. Stems twining, young parts densely yellow villous, later glabrous. Leaf lamina broadly ovate, up to 25·5 × 20·5 cm., acuminate at the apex, rounded or slightly cordate at the base, covered with appressed yellow hairs beneath and on the nerves above; petiole 2·5–9·5 cm. long, yellow-villous. Inflorescence cymose, few-flowered; peduncles up to 20 cm. long, yellow-villous; pedicels up to 2 cm. long, yellow-villous; bracts linear-lanceolate, up to 0·6 cm. long. Sepals linear-lanceolate, subulate, 1·1–1·6 cm. long, the inner ones smaller than the outer ones, herbaceous, hairy on the backs. Corolla funnel-shaped, rose, up to 6·5 cm. long, hairy. Capsule immature globose, brown, coriaceous. Seeds (immature) covered with long fulvous hairs.

   Mozambique. MS: Cheringoma, Chiniziua, nr. Macalaua R., fl. 9.v.1957, *Gomes e Sousa* 4382 (COI; K; LMA; PRE).
   Also in Tanzania. Open forest, bushland and sandy soils.

18. TURBINA Raf.

Turbina Raf., Fl. Tellur. **4**: 81 (1838).
Legendrea Webb & Berth. Hist. Nat. Iles Canar., Bot. **3** (2): 26 (1844).

A scarcely natural genus, similar to *Ipomoea* but differing in having an indehiscent ovoid-oblong or ellipsoid fruit, mostly with a single puberulous seed.

Turbina corymbosa (L.) Raf. has been cultivated in Mozambique, at the Jardim Vasco da Gama of Maputo City (*Balsinhas* 1903), but as far as is known has never become naturalised as it has in other parts of the tropics.

1. Leaf lamina usually oblong or elliptic, covered with strigose yellowish hairs   1. *oblongata*
— Leaf lamina usually ovate-cordate, glabrous or with a pubescence not as above · ·   2
2. Corolla funnel-shaped · · · · · · · · · · 2. *holubii*
— Corolla hypocrateriform · · · · · · · · · · 3
3. Leaves glabrous or more or less pubescent, ovate; outer sepals more or less 10 mm. long, inner ones more or less 13 mm. long · · · · · 3. *stenosiphon*
— Leaves always glabrous, ovate-triangulate, very thin; outer sepals more or less 21 mm. long, inner ones more or less 25 mm. long · · · · · · 4. *longiflora*

1. **Turbina oblongata** (E. Mey. ex Choisy) Meeuse in Bothalia **6**: 778 (1958).—Roessler in Merxm. Prodr. Fl. SW. Afr. **116** 23 (1967).—Ross, Fl. Natal: 296 (1972).—Compton, Fl. Swaziland: 479 (1976). Type from S. Africa.

    Ipomoea oblongata E. Mey ex Drege, Zwei Pfl. Docum.: 46, 142 (1843) *nom. nudum*.
    Ipomoea oblongata E. Mey. ex Choisy in DC., Prodr. **9**: 368 (1845).—Hall. f. in Engl. Bot. Jahrb. **18**: 127 (1893).—Baker & Wright in F.C. **4**: 57 (1904).
    Ipomoea oblongata E. Mey. ex Drege var. *hirsuta* Rendle in Journ. Bot., **39**: 16 (1901). Types from S. Africa (Transvaal).
    Ipomoea lambtoniana Rendle in Journ. Bot. **39**: 16 (1901).—Baker & Wright in F.C. **4**: 61 (1904). Type from Natal.
    Ipomoea randii Rendle in Journ. Bot. **39**: 18 (1901).—Baker & Rendle in F.T.A. **4**: 146 (1905).—Eyles in Trans. Roy. Soc. S. Afr. **5**: 455 (1916). Type: Zimbabwe, Bulawayo, *Rand* 271 (BM, holotype; GRA, isotype).
    Ipomoea seineri Pilger in Engl., Bot. Jahrb. **41**: 297 (1908). Type from Namibia.

Perennial forming several to many annual prostrate or, when still young, suberect stems from a large thick fusiform tuberous root. Stems up to 2 m. long, often suffruticose at base, occasionally thinner, thinly or occasionally densely pubescent with stiff yellowish or brownish hairs, like the petiole, leaf laminas, peduncles and calyces, very rarely glabrous. Leaf lamina very variable in size and shape, usually oblong or elliptic, varying to ovate or linear 2–10 × 1–9 cm., entire with usually rounded, truncate or subcordate base, sometimes broadly cuneate or cordate, obtuse or mucronate, sometimes emarginate, acute or broadly rounded at the apex, with the margin ciliate to sometimes densely ciliate upper surface thinly covered with strigose usually yellowish hairs, lower surface as thinly as the upper or more densely so; petioles usually much shorter than the laminas but occasionally about as long. Inflorescences 1-flowered, sometimes 2-flowered, rarely 3- or 4-flowered; peduncle terete, usually shorter than the leaves; bracteoles very variable but usually lanceolate, acute, hairy like the calyx, sometimes broadly oblong; pedicels usually very short, rarely exceeding 6 mm. Sepals generally lanceolate to ovate-lanceolate, 15–22 mm. long, subequal (inner ones slightly wider) but more or less unequal in specimens with broader oblong or ovate sepals in which the inner ones are narrower; usually acute or acuminate with very acute tips, rarely subobtuse; outer ones more or less densely covered with usually stiff yellowish hairs. Corolla magenta, funnel-shaped 3·5–6 cm. long; midpetaline areas usually thinly covered with silky appressed hairs, rarely quite glabrous. Fruits only rarely produced, subglobose, glabrous, dark brown abruptly apiculate and the apiculus crowned by the persistent style base, enclosed by the slightly accrescent and coriaceous sepals, 1–4-seeded. Seeds glabrous, grey, finely punctate to smooth.

Botswana. N: Between Bushman Pits and Karyer, nr. Nxai Pan, fl. 22.ii.1966, *Drummond* 8840 (SRGH). SW: Ghanzi, c. 42 km. W. of Ghanzi Farm, fl. 7.ii.1971, *van Rensberg* 4204 (PRE). SE: Kanye, 1895, *Marloth* 2156 (PRE). **Zimbabwe**. C: Marondera, fl. 14.i.1942, *Dehn* 562 (SRGH). E: Mutare, Rusambo R., Zimunya Reserve, fl. 8.iv.1962, *Chase* 7679 (K; LISC; SRGH). **Mozambique**. MS: Makurupini Forest, fl. 6.i.1969, *Bisset* 8 (K; SRGH). M: Namaacha, Changalane, Estatuene, 527 m., fl. 9.xi.1967, *Balsinhas* 1147 (LISC; LMA).

    Also in Namibia and S. Africa. Woodland, savanna, grassland, wet and sandy soils; 400–1220 m.

2. **Turbina holubii** (Baker) Meeuse in Bothalia **6**: 780 (1958).—White, F.F.N.R.: 363 (1962).—Roessler in Merxm. Prodr. Fl. SW. Afr. **116**: 23 (1967).—Binns, H.C.L.M.: 41 (1968).—Drummond in Kirkia **10**, 1: 271 (1975). Type: Botswana, Leshumo Valley, *Holub* 512 (K, holotype).

    Ipomoea holubii Baker in Kew Bull. **1894**: 72 (1894).—Baker & Rendle in F.T.A. **4**: 188 (1905).—Eyles in Trans. Roy. Soc. S. Afr. **5**: 454 (1916). Type as above.
    Ipomoea rhodesiana Rendle in Journ. Bot. **39**: 57 (1901).—Baker & Rendle in F.T.A.

4: 188 (1905).—Eyles in Trans. Roy. Soc. S. Afr. **5**: 455 (1916). Type: Zimbabwe, Bulawayo, *Rand* 141 (BM, holotype).

Rivea holubii (Baker) Hall. f. in Meded. Rijksherb. Leiden **1**: 25 (1910). Type as for *Ipomoea holubii*.

Shrub up to 2·5 m. high, many-stemmed from the base and much branched. Stems covered with a light grey bark, slender, sinuous, erect, decumbent or climbing at the tips. Leaves deciduous; leaf lamina ovate-cordate to circular-cordate, sometimes oblong cordate, 1–5·5 × 0·8–4 cm., entire, often folded along the midrib, usually with gradually deflexed, obtuse or subacute, rarely acuminate, often mucronate apex, usually widely and shallowly cordate, sometimes obtuse, truncate or rounded, rarely subacute at the base, more or less sericeous to tomentose on both surfaces, more densely so and more silvery so beneath, sometimes fulvo-sericeous above and on the prominent curving lateral veins beneath, very rarely almost glabrous; petioles rather slender, pubescent like the stem, 4–15 mm. long. Inflorescences 1–5-flowered, axillary, sometimes forming a sort of leaf panicle at the ends of the branches; peduncles slender, sometimes nearly obsolete, usually 2–6 cm. long, hairy like stems, petioles, bracteoles and pedicels; bracteoles elliptic or spathulate to linear oblong, rather small to foliaceous, 2·5 × 6 mm. long, often numerous and forming a sort of involucre at the base of the subumbellate cyme and in this case one of them much larger, occasionally resembling a leaf and up to 3 × 2 cm.; pedicels 6–30 mm. long. Sepals elliptic to obovate or orbicular to obovate-spathulate, 5–12 mm. long, obtuse and mucronate or apiculate, much imbricate, greyish-pubescent to tomentose (at least the outer sepals), sometimes only hairy at the base and conspicuously ciliate, the two inner ones conspicuously larger than the outer ones, all accrescent in fruit, becoming sub-coriaceous-chartaceous, glabrescent, brown, up to about 16 mm. long. Corolla funnel-shaped, pale mauve or pinkish with magenta centre, up to 7·5 cm. long, with a horizontally spreading limb up to about 6 cm. in diam.; midpetaline areas thinly strigose with rather long appressed hairs outside. Stamens very unequal. Ovary glabrous. Fruit ellipsoid, apiculate, with thinly leathery pericarp, usually 1-locular, 1-seeded. Seed ellipsoid, pale yellowish brown or light brown, glabrous, very finely areolate.

Botswana. N: c. 0·8 km. from Francistown, 900 m., fl. 7.iii.1961, *Richards* 14548 (K). **Zambia**. C: Lusaka, Kabulonga, 1220 m., fl. 27.i.1963, *Best* 361 (SRGH). E: Nr. Katete on Petauke side, fl. 26.i.1969, *Anton-Smith* s.n. (SRGH). S: Livingstone, fl. & fr. 3.ii.1961, *Fanshawe* 6175 (FHO; K). **Zimbabwe**. N: Umvukwes, Mutorashanga Pass, 1433 m., fl. & fr. 26.ii.1961, *Leach* 10734 (K; SRGH). W: Hwange National Park, 6 km. from Shumba Pans, SW. of Main Camp, fl. & fr. 18.iv.1972, *Grosvenor* 717 (K; LISC; SRGH). C: Shurugwe c. 16 km. S. of Shurugwe, 1220 m., fl. 24.xii.1959, *Leach* 9679 (SRGH). E: Chimanimani, nr. Moosgwe, fl. & fr. 12.iii.1962, *Plowes* 2237 (K; SRGH). S: c. 27 km. S. of Masvingo, 1070 m fl. & fr., 12.i.1961, *Leach* 10709 (K; SRGH). **Malawi**. N: Rumphi Distr., 30 km. W. of Rumphi on rd. Katumbi, 1250 m., fl. & fr. 12.v.1970, *Brummitt* 10642 (K; SRGH). C: Kasungu Chipala Hill c. 6·4 km. N. of Kasungu, 1000 m., fl. & fr. 14.i.1959, *Robson* 1174 (BM; K; LISC; SRGH). **Mozambique**. N: Cabo Delgado, nr. Mecufi, fl. 3.x.1948, *Pedro & Pedrogão* 5456 (LMA). T: Tete, Chiringa, fl. & fr. 11.iv.1972, *Macedo* 5182 (K; LISC; LMA; LMU; SRGH).

Also in Namibia and S. Africa (Transvaal). Woodland, bushland, grassland, roadside and rocky soil; 900–1500 m.

3. **Turbina stenosiphon** (Hall. f.) Meeuse in Bothalia **6**: 783 (1958).—Verdc. in F.T.E.A., Convolvulaceae: 152, fig. 26 (1963).—Binns H.C.L.M.: 41 (1968).—Drummond in Kirkia **10**, 1: 271 (1975). TAB. **32**. Type from Kenya.

Ipomoea stenosiphon Hall. f. in Sitz.–Ber. Math.–Nat. Akad. Wiss. Wien: 107, Abt. **1**: 50 (1898).—Baker & Rendle in F.T.A. **4**: 192 (1905).—Eyles in Trans. Roy. Soc. S. Afr. **5**: 455 (1916).—Brenan, T.T.C.L. 172 (1949).—White, F.F.N.R.: 362 (1962). Type as above.

Rivea stenosiphon (Hall. f.) Hall. f. in Jahrb. Hamb. Wiss. Anst. 16, Beih. **3**: 15 (1898).

Turbina stenosiphon (Hall. f.) Meeuse var. *pubescence* Verdc. in Kew Bull. **13**: 217 (1958). Type from Tanzania.

A tall woody climbing shrub up to 10 m. Stems terete, glabrous, more or less robust, flowering branches densely leafy especially towards the apex and frequently with rugose bark. Leaves sometimes fascicled on very short branchlets; leaf lamina ovate, 6–11 × 4–9 cm., obtusely acute or emarginate at the mucronate apex, cordate at the base, glabrous or more or less pubescent, mainly on the nerves beneath, entire or more or less

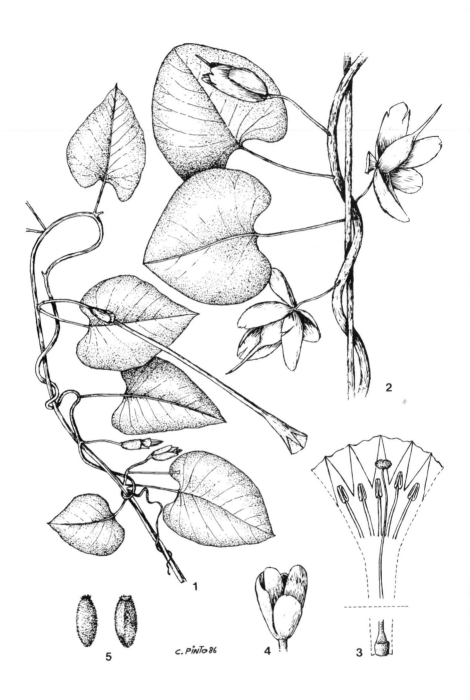

Tab. 32. TURBINA STENOSIPHON. 1, flowering branch (×½), from *Fanshawe* 8408; 2, fruiting branch (×½), from *Pope* 232; 3, corolla opened to show stamens and pistil (×1); 4, calyx (×1); 5, seed (×1), 3–5 from *Fanshawe* 8408.

subrepand; venation reticulate beneath; lower surface dotted with numerous small circular glands; petiole slender 2–9 cm. long. Flowers said to be evil-smelling, opening at night, fascicled on short shoots; peduncles very short or obsolete, 0–3 mm. long articulated against the solitary or 2–5 fascicled, glabrous; pedicels, 2–6 cm. long; bracts very early deciduous, more or less membranous, yellowish-brown, oblong-lanceolate, 2–3 mm. long, glabrous, more or less acute. Sepals unequal, elliptic obtuse, glabrous or slightly pubescent, thinly coriaceous, the outer ones up to 10 mm. long and the inner ones up to 13 mm. long, all in fruit becoming much enlarged. Corolla white, hypocrateriform; the tube very slender up to 15 cm. long; limb salver-shaped about 5·5 cm. wide, with woolly hairs near the apices of the midpetaline areas outside. Stamens and styles long exserted. Capsule ellipsoidal, dirty brown, tipped by the persistent 1·5 cm. long style-base, 1(3)-seeded; fruiting. Sepals up to 4 × 2 cm. Seeds purple-brown, velvety with short hairs.

Zambia. W: Kitwe, st. 29.viii.1959, *Fanshawe* 5142 (K). E: Chipata, Machinje Hills, st. 15.v.1965, *Mitchell* 9253 (K). **Zimbabwe**. N: Darwendale, Great Dyke, fl. & fr. 26.iii.1963, *Wild* 6090 (K; LISC; SRGH). W: Bulawayo, Cyrene Mission, fl. 11.i.1972, *Ewbank* in GHS 221734 (K; LISC; SRGH). E: Mutare, Zimunya's Reserve, 915 m., fr. 1.vi.1958, *Chase* 6929 (COI; K; LISC; SRGH). S: Gutu, between Mvuma and Masvingo, c. 50 km. from Masvingo, fr. 2.v.1970, *Pope* 232 (K; LISC; SRGH). **Malawi**. N: Mzimba. Phopo Hill, NE. end of Lake Kazuni, 110–1150 m., fl. & fr. 20.v.1970, *Brummitt* 10958 (EA; K; LISC; MAL; PRE; SRGH; UPS). C: C. 32 km. S. of Lilongwe, fr. 7.vi.1938, *Pole Evans & Erens* 612 (K; PRE). **Mozambique**. MS: Manica, between Rotland and Mavita, fl. & fr. 17.vi.1949, *Pedro & Pedrogão* 6639 (LMA).
Also in tropical Africa and S. Africa (Transvaal). Woodland, river banks, boulders and roadsides; 1000–1650 m.

4. **Turbina longiflora** Verdc. in Kirkia **6**: 121, photo 5 (1967). Type: Mozambique, Maputo, Goba, nr. Fonte dos Libombos, *Barbosa & Lemos* 8271 (LMA, holotype; K, isotype).

Perennial, glabrous herb, climbing. Stems slender, more or less ridged, yellowish. Leaf lamina ovate-triangulate, 2·8–6 × 2–4·5 cm., cordate-emarginate at the base, apex subacute or rounded, mucronate, glabrous, densely dotted with small circular glands, entire or more or less subrepand, very thin; petiole very slender 2·7–8 cm. long. Flowers axillary, solitary or in 1–2 on lateral abbreviate branches; peduncle up to 10 mm. slender; pedicels 3 cm. long, slender, thickened at the apex. Sepals unequal; the 2 outer ones linear-oblong or elliptic-oblong up to 21 mm. long, the 3 inner ones linear up to 25 mm. long, all rounded at the apex and glabrous. Corolla white, hypocrateriform; the tube very slender up to 12 cm. long; limb salver-shaped up to about 4 cm. wide. Stamens and style exserted. Disk 4-lobed. Fruit not seen.

Mozambique. N: Macondes, between Nantulo and Mueda, 310 m., fl. 10.iv.1964, *Torre & Paiva* 11854 (C; LISC; LMA; WAG). GI: Inhambane, Govuro, between Banamana and Machaíla, fl. 26.iii.1974, *Correia & Marques* 4208 (LISC; LMU). M: Namaacha, Goba, nr. Fonte dos Libombos, fl. 27.iii.1958, *Barbosa & Lemos* 8271 (COI; K; LISC; LMA, holotype; SRGH).
Not known elsewhere. Woodland, sandy soils; 310 m.

Argyreia nervosa (Burm. f.) Boj., a climber up to 10 m. high, densely whitish or fulvous tomentose, with large ovate or orbicular cordate leaves and pink-purple flowers c. 6 cm. long, is native in India from Assam and Bengal to Belgaum and Mysore and cultivated as a garden plant in Mozambique.

19. PARALEPISTEMON Lejoly & Lisowski

Paralepistemon Lejoly & Lisowski in Bull. Jard. Bot. Nat. Belg. **56**: 196 (1986).

This genus differs from *Ipomoea* as follows: Stamens inserted on triangular, large (more or less 2 × 2 mm.) glandular pilose scales situated near the base of the corolla tube. Ovary with a very short beak persistent in fruit. Style articulated at the base, caducous. Fruit indehiscent, more or less woody.

Paralepistemon shirensis (Oliv.) Lejoly & Lisowski in Bull. Jard. Bot. Nat. Belg. 56: 197 (1986). TAB. **33**. Type: Malawi, Shire Highlands, *Buchanan* 262 (K, lectotype).
    Ipomoea shirensis Oliv. in Hook., Ic. Pl., Ser. III, **5**: 58, t. 1474 (1884).—Baker & Rendle in F.T.A. **4**: 189 (1905). Type as above.

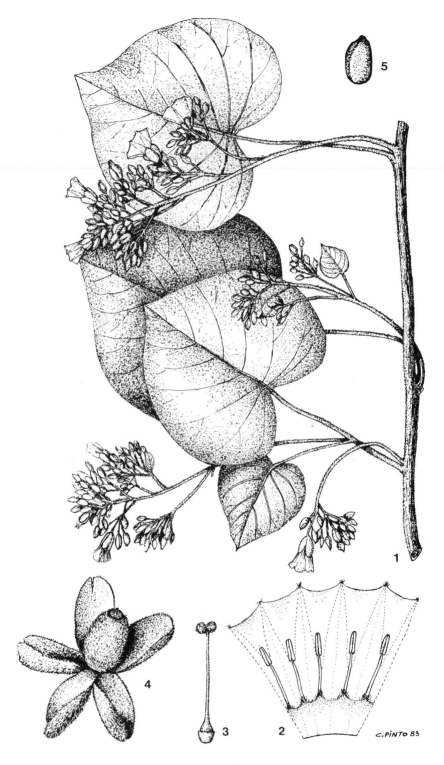

Tab. 33. PARALEPISTEMON SHIRENSIS. 1, habit (×½); 2, corolla opened to show stamens (×2); 3, pistil with nectary disk (×2); 4, fruit (×2); 5, seed (×2). All from *Lemos & Macuàcua* 128.

Rivea shirensis (Oliv.) Hall. f. in Engl., Bot., Jahrb. **18**: 157 (1893); Jahrb. Hamburg. Wissensch. Anst. 16, Beih. **3**: 14 (1898). Type as above.

Porana subrotundifolia De Wild., Et. Fl. Katanga in Ann. Mus. Congo, Bot. Ser. **1**: 111, t. 5, fig. 8–18 (1902–3). Types from Zaire.

Turbina shirensis (Oliv.) Meeuse in Bothalia **6**: 782 (1958).—White, F.F.N.R.: 363 (1962).—Binns, H.C.L.M.: 41 (1968).—Jacobsen, Check-List Fl. Lomagundi Distr. Rhod.: 171 (1973).—Drummond, List Trees, shrubs and woody climbers indigenous naturalised Rhod.: 271 (1975). Type as for *Paralepistemon shirensis*.

A tall robust climber up to 15 m. tall. Stems woody, terete, up to 20 cm. in diam., whitish tomentose of very short appressed hairs, more densely so when young. Leaf lamina broadly cordate to circular-ovate or cordate-ovate, entire, obtuse to acuminate or shortly and abruptly cuspidate at the apex, often mucronulate, 3·5–17 × 3–16 cm. the basal sinus wide and shallow to leaf base almost truncate, upper surface thinly pubescent, more densely so when young, densely white or greyish tomentose beneath; petioles rather slender 1·5–12 cm. long, densely tomentose. Inflorescences cymose in the axils of the upper leaves and forming a lax terminal panicle, sometimes pseudo-umbellate; peduncles patent or erecto-patent, rather slender, densely tomentose like the young stems and petioles, few–many-flowered, 3–10 cm. long; bracteoles thinly papery or almost membranous, oblong-oblanceolate, oblong-obovate or somewhat narrowly oblong-spathulate, much narrowed and subpetiolate at the base, 7–22 mm. long, pale yellowish brown when dry hairy outside, glabrous inside, very early deciduous and rarely preserved; pedicels up to 2 cm. long, densely tomentose. Sepals thin, almost papery, subequal, oblong or obovate-oblong, obtuse, densely sericeo-tomentose outside, 8–15 mm. long, at first erect, much imbricate, later accrescent, glabrescent and much spreading, ultimately papery, brown and often purplish outside, pale straw-coloured inside, 10–17 mm. long. Corolla white (not lilac as stated in F.T.A.) widely funnel-shaped, 15–20 mm. long; midpetaline areas sharply defined, densely silky. Fruit indehiscent broadly ellipsoid or somewhat obovoid, rounded-truncate to depressed at the apex and shortly beaked, very densely greyish sericeo-tomentose, 1-seeded, with a hard woody pericarp. Seed light brown or fawnish, subglobose-quadrangular, very shortly velutinous or puberulous.

Zambia. W: Luswishi/Ndola, fl. & fr. 30.iii.1979, *Chisumpa* 519 (K). C: Mt. Makulu Res. Sta. nr. Chilanga, fl. & fr. 1.v.1960, *Angus* 2232 (FHO; K; SRGH). E: Luangwa Valley, Saili Munkanya, fl. & fr. 9.iv.1968, *Phiri* 142 (K). S: Mazabuka Res. Sta., fl. & fr. 21.v.1965, *Lawton* 1213 (K; SRGH). **Zimbabwe**. N: Urungwe, Kariba Gorge, Rutanswa R., fl. & fr. 17.iv.1955, *Lovemore* 432 (COI; K; SRGH). W: Hwange, c. 11 km. from Bulawayo-Victoria Falls Rd., fl. & fr. 1.iv.1975, *Raymond* 320 (SRGH). E: Mutare, Hondi Valley, fr. vii.1949, *Chase* 1709 (BM; K; SRGH). S: Gwanda, Doddieburn Ranch, Makoli Kopje, c. 840 m., fl. & fr. 11.v.1972, *Pope* 751 (K; SRGH). *Malawi*. S: Chikwawa, 760 m., fl. & fr. 17.v.1949, *Gerstner* 7066 (K; PRE; SRGH). **Mozambique** N: c. 5 km. E. of Malema, 670 m., fl. & fr. 16.v.1961, *Leach & Rutherford-Smith* 10873 (K; SRGH). Z: Mocuba, Namagoa, 200 k. inland from Quelimane, 60–120 m., fl. & fr. v.1943, *Faulkner* 96 (K; PRE). T: Tete, Songo, c. 860 m., fl. 3.iv.1972, *Macedo* 5140 (K; LISC; LMA; LMU; SRGH). MS: Chemba, Chiou, Exp. Sta. CICA, fl. 21.iv.1960, *Lemos & Macuácua* 122 (BM; COI; K; LISC; LMA; SRGH). GI: Guijá, Caniçado, Chirunzo, fl. & fr. 3.vii.1947, *Pedro & Pedrogão* 1261 (COI; K; LMA; SRGH).

Also in Zaire, Angola and S. Africa (Transvaal). Woodland, dry riverine forest, grassland, sandy soils and roadsides; 60–1100 m.

CUSCUTACEAE

By Maria Leonor Gonçalves

Twining parasitic, usually glabrous herbs, almost without chlorophyl, annual or rarely perennial in the tissues of the host, attached by means of numerous haustoria; free-living for a brief period after germination until attachment to the host is accomplished; hairs, when present, mostly unicellular or bicellular, not glandular. Stems usually terete and slender to filiform, often whitish, yellowish or reddish; the vascular system somewhat reduced, without internal phloem, and vessel segments with simple perforations. Leaves reduced to minute scales or absent. Inflorescences lax or compact cymose clusters. Flowers small, sessile or shortly pedicelled, 5-merous or less often 4-merous, rarely 3-merous, as to the calyx, corolla and androecium. Calyx lobed or parted with sepals united at the base, each one with a single vascular trace. Corolla usually white or pink, lobed; lobes united into a tube at the base, shorter or longer than the tube, often patent or reflexed; the tube inside usually with a whorl of thin fringed scales opposite to and below the stamens. Stamens inserted at the throat and alternating with the corolla-lobes; filaments often short; anthers often broadly elliptic, tetrasporangiate and dithecal, opening by longitudinal slits; pollen smooth. Ovary 2(3)-locular, each loculus with 2 ovules, erect on basal axile (or intruded-parietal) placentas, anatropous; base of the ovary at least sometimes nectariferous; styles 2, terminal, distinct or connate into a single column; stigmas capitate or linear. Fruit an ovoid or subglobose capsule, opening irregularly, or circumscissile near the base, or indehiscent. Seeds 4 or less, subglobose or angular, often granular, almost invariably glabrous; embryo scarcely differentiated, straight, filiform, sometimes enlarged at one end; cotyledons absent or rudimentary.

A monogeneric family sometimes included in the *Convolvulaceae* but distinct from this because the different habit and being parasitic, usually without chlorophyll, without glandular hairs and the vascular system reduced; leaves scale-like or none; sepals with a single vascular trace; corolla mostly with hypostaminal scales present; pollen smooth; embryo scarcely differentiated.

CUSCUTA

Cuscuta L., Sp. Pl. 1: 124 1753); Gen Pl. ed. 5: 60 (1754).

Characters as for the family.

A genus of about 150 species, almost cosmopolitan but best developed in the new world, especially in warmer regions.

1. Styles almost to the apex united into a single column · · · · *6. cassytoides*
— Styles free to the base or nearly so · · · · · · · · · 2
2. Stigmas capitate-globose or more or less peltate · · · · · 3
— Stigmas linear · · · · · · · · · *7. planiflora*
3. Stigmas (in dried specimens) sometimes more or less peltate with convolute edges; flowers c. 4–7. long and broad more or less; calyx and corolla lobes broad, obtuse or rounded; capsule vertically circumscissile · · · · · · *4. kilimanjari*
— Stigmas usually globose; flowers c. 2–4 mm. long and broad; capsule not circumscissile or if so then corolla-lobes very acute · · · · · · · · 4
4. Corolla-lobes obtuse · · · · · · · *1. australis*
— Corolla-lobes acute · · · · · · · · · 5
5. Scales absent; calyx-lobes acute · · · · · · *5. hyalina*
— Scales present · · · · · · · · 6
6. Inflorescence a compact spherical cluster; calyx-lobes obtuse; scales usually reaching the base of the filaments · · · · · · · *2. campestris*
— Inflorescence a very loose cluster; calyx-lobes obtuse or abruptly acute; scales usually not reaching the base of the filaments · · · · · · *3. suaveolens*

1. **Cuscuta australis** R. Br., Prodr. Fl. Nov. Holl.: 491 (1810).—Yuncker in Mem. Torr. Bot.

Club. **18**: 124, fig. 1 (1932).—Meeuse in Bothalia **6**: 647 (1958).—Heine in F.W.T.A. ed. 2,
2: 336 (1963).—Verdc. in F.T.E.A., Convolvulaceae: 4 (1963).—Binns, H.C.L.M.: 39
(1968). Type from Australia.

Cuscuta obtusiflora H.B.K., var. *cordofana* Engelm. in Trans. Acad. Sci. St. Louis **1**: 493
(1859).—Dammer in Engl., Pflanzenw. Ost-Afr. **C**: 334 (1895).—Rendle in F.T.A. **4**; 204
(1906) pro parte.—Eyles in Trans. Roy. Soc. S. Afr. **5**: 452 (1916). Type from the Sudan.

Cuscuta obtusiflora H.B.K. var. *australis* (R.Br.) Engelm. in Trans. Acad. Sci. St. Louis
1; 492 (1859).

Cuscuta cordofana (Engelm.) Yunker in Mem. Torr. Bot. Club **18**: 127, fig. 2 (1932).

Cuscuta chinensis sensu Hutch. & Dalz., F.W.T.A. **2**: 219 (1931) non Lam.

Stems medium thick up to 0·5 mm. in diam. Inflorescences rather loose to dense
cymose and subglobose clusters. Flowers 2–3 mm. long and broad, accrescent in fruit,
often somewhat glandular, subsessile, often obpyriform in fruit. Calyx about as long as
the corolla tube; lobes broad, ovate-circular, obtuse, not overlapping. Corolla campanulate; lobes rounded, obtuse or somewhat triangular but blunt, shorter than or as
long as the tube, erect to somewhat spreading. Stamens shorter than the corolla lobes;
filaments usually very short and stout, more or less subulate; anthers broadly elliptic to
subcircular. Scales shorter than the tube, reaching nearly to the base of the filaments,
much fringed apically, not bifid or variously bifid sometimes consisting of one fringed
lobe on either side of the area of the corolla tube below the filaments. Ovary globose;
styles stout, shorter than the ovary, divergent and intrastylar aperture large. Capsule
globose or somewhat obpyriform, not splitting at base; intrastylar opening 1–1·5 mm.
long, often almost circular. Seeds ellipsoid to ovoid up to 1·5 mm. long.

Botswana. N. Okovango delta, Thaoge R., fl & fr. 24.viii.1975, *Smith* 1454 (K; SRGH).
Zambia. C: Kabwe Distr., Mufukushi R., near Chipepo, 1130 m., fl & fr. 20.i.1973, *Kornas*
3049 (K). S: 10 km. W. of Kasha, 1020 m., fl. & fr. 20.iv.1965, *Seymoens* 11514 (K).

Widespread from the Sudan Republic to central and western Africa, S. Africa and Madagascar;
throughout Old World from southern Europe to Japan and Australia. On swamp vegetation, e.g.
Leersia, *Nymphaea*, *Papyrus* and *Polygonum*; 900–1200 m.

2. **Cuscuta campestris** Yunker in Mem. Torr. Bot. Club **18**: 138, fig. 14 (1932).—Meeuse in
Bothalia **6**: 648 (1958).—Heine in F.W.T.A. ed. 2, **2**: 336 (1963).—Verdc. in F.T.E.A.,
Convolvulaceae: 5 (1963).—Ross, Fl. Natal: 294 (1972).—Compton, Fl. Swaziland: 473
(1976). TAB. **34**: Type from the U.S.A.

Stems slender, up to 0.3 mm. in diam. Inflorescences very dense cymose globose
clusters. Flowers up to 2 mm. long and broad, greenish-yellow, often glandular;
pedicels usually shorter than the flowers. Calyx about enclosing the corolla tube,
broadly campanulate; lobes ovate to circular or broadly triangular, usually obtuse to
rounded, mostly overlapping when young. Corolla campanulate; lobes triangular,
acute, equalling the corolla tube, spreading with often inflexed apices. Stamens shorter
than corolla lobes; filaments longer than, or about equalling, the anthers. Scales large,
adnate only at base, much-fringed apically and almost or quite reaching the anthers.
Ovary globose; styles slender, becoming thicker and conspicuous in fruit. Capsule
usually depressed-globose with a depression around the style bases, not splitting at the
base, with the intrastylar opening small. Seeds ovate, about 1 mm. long, flattened on
one side.

Botswana. N: Ngamiland Distr., Ngokha R., Okavango Swamp, fl. & fr. 16.ii.1973, *Smith*
396 (K; SRGH). SE: Mahalapye Distr., Sefare, fl. & fr. 25.ii.1958, *Beer* 692 (SRGH). **Zambia**.
W: Kitwe, Mwambashi, 20.xi.1970, *Fanshawe* 10994 (K; SRGH). C: Chalimbana R. near
Liempe, 18 km. E. of Lusaka, 1200 m., fl. & fr 5.i.1972, *Kornas* 792 (K). S: Choma, 4300 m.,
28.iii.1955, *Robinson* 1221 (K). **Zimbabwe**. N: Mazoe, Marodzi R., Selkirk Farm, Uvinga, fl. &
fr 23.iii.1970, *Townsend* 23088 (LISC; SRGH). W: Bulawayo, Lower Umgusa irrigation
scheme, fl. & fr. 5.xii.1958, *Laan* 92989 (K; SRGH). C: Harare, fl. & fr. 23.ii.1971, *Biegel* 3639
(K; SRGH). E: Mutare, La Rochelle, Imbeza Valley, 1190 m., fl. & fr. 13.i.1964, *Chase* 8102 (K;
SRGH). **Mozambique**. M: Marracuene, Movedja, Incomati Valley, fl. & fr. 3.ix.1961,
Macuácua 96 (BM; K LMA; LISC; LMA; SRGH).

A native of America, widely naturalised in the Old World.

Parasitic on swamp vegetation and on a great variety of other plants, some in gardens;
1000–4300 m.

3. **Cuscuta suaveolens** Ser. in Ann. Sci. Phys. Nat. Agric. & Indust. **3**: 519 (1840).—Yunker
in Mem. Torr. Bot. Club **18**: 148, fig. 22 (1932).—Wild, Common Rhod. Weeds fig. 92

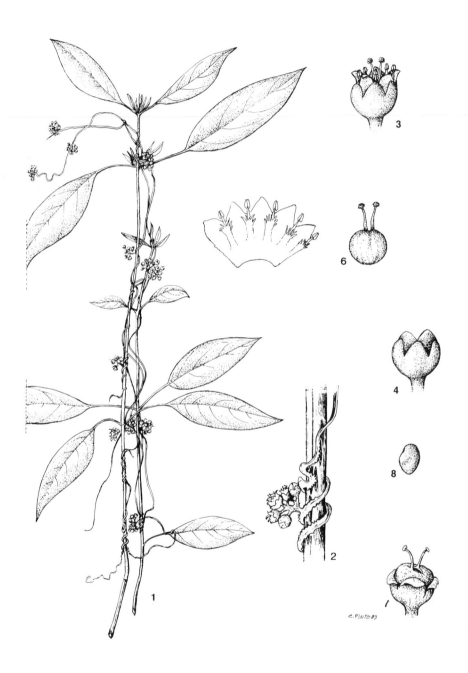

Tab. 34. CUSCUTA CAMPESTRIS. 1, habit ($\times\frac{1}{2}$); 2, part of plant affixed to the host by means of haustoria ($\times 2\frac{1}{2}$); 3, flower ($\times 6$); 4, calyx ($\times 6$); 5, corolla opened to show stamens and scales ($\times 6$); 6, pistil ($\times 6$); 7, fruit enclosed in calyx ($\times 6$); 8, seed ($\times 6$). All from *Townsend* in GHS 203088.

(1955).—Meeuse in Bothalia **6**: 648 (1958).—Verdc. in F.T.E.A., Convolvulaceae: 5 (1963). Type from France, cult. at Lyons from seeds probably from Chili.

Stems slender to medium thick. Inflorescences loose racemose clusters. Flowers 3–4 mm. long, more or less glandular; pedicels variable in length. Calyx shorter than the corolla tube; lobes ovate-triangular, more or less acute, not overlapping, often with revolute edges, separated by usually rounded sinuses. Corolla campanulate or funnel-shaped, becoming globular about the developing capsule; lobes ovate-triangular with acute inflexed and somewhat thickened apices shorter than the tube. Stamens shorter than corolla lobes; filaments about as long as the anthers. Scales usually not reaching the stamens, oblong, ovate or triangular-ovate irregularly fringed. Ovary globose; styles long and slender, about equal to the ovary or sometimes longer and distinctly unequal. Capsule globose, not splitting around the base. Seeds subglobose, 1·5–2 mm. long.

Zimbabwe. C: Harare, Agriculture Exp. Station, fl. & fr. 2.ii.1936, *Bacon* 6510 (K; SRGH). Native of S. America, widespread and now a cosmopolitain weed. Parasitic usually on Lucerne (*Medicago sativa*).

4. **Cuscuta kilimanjari** Oliv. in Johnston Kilimanjaro Exp. Append.: 343 (1886), nom. nud.; Trans. Linn, Soc. Bot., Ser. 2, **2**: 343 (1887).—Baker & Rendle in F.T.A. **4**: 205 (1906).—Rendle in Journ. Linn. Soc. Bot. **40**: 150 (1911).— Eyles in Trans. Roy. Soc. S. Afr. **5**: 452 (1916).—Yuncker in Mem. Torr. Bot. Club **18**: 187, fig. 58 (1932).—Dandy in Andr., Fl. Pl. Anglo-Egypt. Sudan **3**: 109 (1956).—Meeuse in Bothalia **6**: 650 (1958). —Verdc. in F.T.E.A., Convolvulaceae: 6, fig. 1 (1963).—Binns, H.C.L.M.: 39 (1968). —Jacobsen in Kirkia **9**: 171 (1973). Type from Tanzania.

Var. **kilimanjari**
 Cuscuta ndorensis Schweinf. in von Höhnel, Zum Rudolph-See und Stephanie-See: (1892). Type from Kenya.
 Cuscuta obtusiflora H.B.K. var. *cordofana* sensu Baker & Rendle in F.T.A. **4**: 204 (1906) pro parte excl. spec. Figari tantum non Engelm.

Stems medium thick to rather stout, up to 1·5 mm. in diam. Inflorescences few-flowered cymes. Flowers 4–6 mm. long and about as broad, pale cream, somewhat coriaceous when dried; pedicels shorter than the flowers. Calyx cupulate nearly enclosing the corolla; lobes ovate-circular, obtuse, overlapping at the base, rather thick and often more or less carinate. Corolla campanulate-cylindric; lobes ovate or circular, obtuse, shorter than the tube, often revolute. Stamens shorter than the corolla lobes; filaments short and thick. Scales triangular or oblong, rather thick, reaching to the base of the stamens, but sometimes smaller, fringing variable in amount. Ovary globose; styles shorter than the ovary and rather thick, often reflexed; stigmas mostly flattened and convolute. Capsule depressed-globose, intrastylar aperture large, at length splitting around the base. Seeds subglobose, up to 2 mm. long, pale yellow-brown or blackish when dry.

Zambia. N: Mbala, Chilongowelo, 1500 m., fl. & fr. 21.iv.1952, *Richards* 1518 (K). E: Nyika, fl. & fr. 26.vi.1966, *Fanshawe* 9748 (SRGH). **Zimbabwe**. N: Chipuriro, Mpingi Pass, Great Dyke, 1370 m., fl. & fr 18.v.1962, *Wild* 5783 (K; LISC; SRGH). E: Chipinge, Chirinda Forest, 1160 m. fl. & fr.20.v.1906, *Swynnerton* 453 (BM; K; SRGH). **Malawi**. N: Rumphi, Chelinda to Zambia Rest House, 2200 m., fl. & fr. ix.1967, *Michael & Hinchely* 77 (SRGH). C: Nandi Forest, *Johnston* s.n. (K). S: Blantyre, Ndirande Mt., 1370–1520 m., fl. & fr. 28.vi.1970, *Brummitt* 11719 (K; SRGH). **Mozambique**. MS: Mossurizie, Espungabera, fl. & fr. 13.vi.1942, *Torre* 4325 (LISC; LMA; WAG).
 Throughout eastern Africa, Ethiopia to S. Africa, mostly above 1000 m. Parasitic on various hosts, mainly shrubby, often on *Acanthaceae* on floors and edges of lowland and upland rain forest.
 This species is readily recognised by its large flowers; the anthers dry a pale cream-colour contrasting with dark brown of the rest of the flower.

5. **Cuscuta hyalina** Heyne ex Roth, Nov. Pl. Sp.: 100 (1821).—Dammer in Engl., Pflanzenw. Ost-Afr. **C**: 334 (1895).—Baker & Rendle in F.T.A. **4**: 205 (1906).—Yuncker in Mem. Torr. Bot. Club **18**: 235, fig. 107 (1932).—Dandy in Andr., Fl. Pl. Anglo Egypt. Sudan **3**: 108 (1956).—Meeuse in Bothalia **6**: 650 (1958).—Verdc. in F.T.E.A., Convolvulaceae: 8 (1963).—Roessler in Merx m. Prodr. Fl. S. Afr. **117**: 1 (1967). Type from India "orientale".
 Cuscuta epitribulum Schinz in Bull. Herb. Boiss., Sér. 2, **1**: 880 (1908).—Baker & Rendle in F.T.A. **4**: 206 (1906). Type from Namibia.

Stems usually very slender, under 0·5 mm. in diam. Inflorescences rather loose umbellate clusters. Flowers 2–3·5 mm. long on short pedicels, thin in texture, sometimes 4-merous, shining and yellowish when dry, often glandular. Calyx campanulate-turbinate; lobes triangular-ovate, acuminate often exceeding and enclosing the corolla, erect to reflexed. Corolla campanulate, becoming globular about the developing fruit; lobes narrow and very acute, erect or spreading and reflexed in old flowers, usually longer than the tube. Stamens shorter than the corolla lobes; filaments somewhat stout, as long or longer than anthers. Scales absent. Ovary globose, more or less depressed; styles slender, as long as or longer than the ovary, somewhat unequal. Capsule depressed-globose, irregularly splitting near the base; intrastylar opening small with a longitudinal groove from each end running down the capsule. Seeds ovoid, about 1·5 m. long.

Botswana. SE: Gaberones, Content Farm, fl. & fr. 15.iii.1973, *Kelaole* 155 (SRGH). **Zimbabwe** E: Chimanimani, Birchenough Bridge, 700 m., fl. & fr. 17.xii.1952, *Chase* 4741 (K; SRGH). S: Bikita, Sabi R. bank, fl. & fr. 11.ii.1966, *Wild* 7530 (K; LISC; SRGH).
India to Ethiopia, the Sudan Republic and drier areas of southern Africa. Parasitic on various hosts in rather dry country; 700–1500 m.

6. **Cuscuta cassytoides** Engelm. in Trans. Acad. Sci. St. Louis **1**: 513 (1859).—Peter in Engl. & Prantl, Pflanzenfam. ed. 1, **4**, 3a: 39 (1891).—Dammer in Engl., Planzenw. Ost-Afr. **C**: 334 (1895).—Baker & Wright in Dyer, F.C. **4**, 2: 86 (1904).—Baker & Rendle in F.T.A. **4**: 206 (1906).—Yuncker in Mem. Torr. Bot. Club **18**: 250, fig. 123 (1932).—Meeuse in Bothalia **6**: 651 (1958).— Verdc. in F.T.E.A., Convolvulaceae: 8 (1963).—Ross, Fl. Natal: 294 (1972).—Compton, Fl. Swaziland: 473 (1976). Type from S. Africa (Cape Prov.).
   Cuscuta timorensis Engelm. in Trans. Acad. Sci. St. Louis **1**: 514 (1859.—Yuncker in Mem. Torr. Bot. Club **18**: 250, fig. 124 (1932).—van Ooststr. in Fl. Males, Ser. 1, **4**: 393 (1953). Type from Timor.

Stems very coarse up to 2.5 mm. in diam., often with purple spots. Inflorescences few-flowered cymes arranged in lax paniculate spikes or reduced to a single short raceme. Flowers up to 4 mm. long, subsessile. Calyx cupulate, almost as long as the corolla tube, lobes ovate-circular, very obtuse, overlapping. Corolla campanulate; lobes as long as the tube, ovate, obtuse to rounded, erect to reflexed. Stamens shorter than the corolla lobes; filaments much shorter than the anthers. Scales triangular adnate to the tube over most of their surface, with small free lateral fringed portions, sometimes much reduced. Ovary globose-conic; styles united into a single column about 0·5(1·5) mm. long; stigmas small, flat. Capsule ovoid-conic, splitting round the base; styles persistent. Seeds subglobose, up to 3 mm. long.

Zambia. N: Mafingi Mts., fl. & fr. 31.x.1972, *Fanshawe* 11643 (K). **Mozambique**. MS: Manica, Mavita, between Revue R. and Munhinga, fl. & fr. 27.iv.1948, *Barbosa* 1601 (LISC; LMA; WAG).
East Africa to South Africa; also Java and Lesser Sunda Is. Parasitic on many woody plants and herbs.

7. **Cuscuta planiflora** Tenore, Fl. Napolit. 3: 250 (1824–9).—Dammer in Engl., Pflanzenw. Ost-Afr. **C**: 334 (1895).—Hiern in Cat. Afr. Pl. Welw. **1**: 743 (1898).—Baker & Rendle in F.T.A. **4**: 203 (1906).—Eyles in Trans. Roy. Soc. S. Afr. **5**: 452 (1916).—Yuncker in Mem. Torr. Bot. club **18**: 292 (1932).—Meeuse in Bothalia **6**: 655 (1958).—Verdc. in F.T.E.A., Convolvulaceae: 9 (1963).—Roessler in Merxm., Prodr. Fl. SW. Afr. **117**: 1 (1967).—Jafri in Fl. Libya **53**: 5 (1978). Type from Italy.

Stems slender, up to 0.3 mm. in diam., yellowish or crimson. Inflorescences more or less compact clusters. Flowers up to 3 mm. long, usually whitish, subsessile, more or less fleshy. Calyx about enclosing the corolla tube or shorter, broadly campanulate; lobes fleshy or turgid, obtuse or acute. Corolla campanulate-globose; lobes more or less acute or slightly obtuse, membranous or turgid at the apices, about equalling the tube or shorter, spreading. Scales usually oblong, about reaching the stems or shorter, fringed, usually bifid, bridged low down or at the middle. Stamens shorter than the corolla lobes, filaments about equalling or longer than the anthers. Ovary globose; styles longer or shorter than the stigmas, slender; stigmas slender. Capsule depressed-globose, markedly splitting around the base. Seeds ovate, granulated, mostly less than 1 mm. long.

Flowers small; calyx more or less enclosing the corolla tube; corolla about 1·5–2·5 mm. long; corolla-lobes abruptly acute - · - · - · - · - · - var. *planiflora*
Flowers larger; calyx shorter than the corolla tube; corolla 2–4 mm. long; corolla lobes acute or rather obtuse - · - · - · - · - · - var. *approximata*

Var. **planiflora**.

Cuscuta planiflora Tenore var. *mossamedensis* Hiern in Cat. Afr. Pl. Welw. **1**: 743 (1898).—Meeuse in Bothalia **6**: 655 (1958). Type from Angola.

Cuscuta balansae Boiss. & Reut. var. *mossamedensis* (Hiern) Yuncker in Mem. Torr. Bot. Club **18**: 291, fig. 154 F-G (1932). Type as above.

Flowers small in small clusters. Calyx up to 3·5 mm. across when flattened out, more or less enclosing the corolla tube. Corolla about 1·5–2·5 mm. long; tube about 1·5 mm. long; lobes abruptly acute. Styles and stigmas together up to 1 mm. long., about equalling one another or styles shorter than the stigmas.

Zambia. N: Mbala, Uningi Pans, 1500 m., fl. & fr. 23.iii.1966, *Richards* 21376 (K; SRGH). E: Katondwe, fl. & fr. 23.ii.1965, *Fanshawe* 9208 (SRGH). S: Namwala, Mufunta Camp, fl. & fr. 19.i.1968, *Hanks* JH/1/68 (SRGH). **Zimbabwe**. N: Mazoe, Iron Mask Hills, 1500 m., iv.1906, *Eyles* 327 (BM; SRGH). W: Hwange, Shapi Camp, 1000 m., fl. & fr. 27.ii.1967, *Rushworth* 273 (K; LISC; SRGH). C: Harare, Parktown, fl. & fr. 23.iv.1944, *Greatrex* 18415 (SRGH).
Widespread throughout North and East Africa to Angola and Namibia. Also in the Mediterranean region and SW. Asia. Parasitic on a great variety of herbaceous plants of mopane woodland, in grassland and seasonally damp and granitic and sandy soil; 1000–1500 m.

Var. **approximata** (Bab.) Engelm. in Trans Acad. Sci St. Louis **1**: 465 (1859).—Verdc. in F.T.E.A., Convolvulaceae: 10 (1963). Type a specimen cultivated in England, from seed from Afghanistan.

Cuscuta approximata Bab. in Ann. Mag. Nat. Hist. **13**: 253 (1844).—Yuncker in Mem. Torr. Bot. Club **18**: 295, fig. 158 (1932). Type as above.

Cuscuta rhodesiana Yuncker in Bull. Torr. Bot. Club **84**: 429 (1957). Type: Zambia, Ndola, *Young* 53 (BM, holotype).

Flowers and clusters larger than in var. *planiflora*. Calyx 3–4·5 mm. across when flattened out, shorter than the corolla tube. Corolla 2–4 mm. long; tube up to 3 mm. long; lobes acute or rather obtuse. Styles and stigmas together about 1–1·3 mm. long; stigmas equalling or twice as long as the styles.

Zambia. N: Mbala, Lumi R. Flats (Dambo), 1500 m., fl. & fr. 17.viii.1956, *Richards* 5856 (K). W: Solwezi, Mwinilunga Rd., 117 km. of Solwezi, 1240 m., fl. & fr. 14.v.1972, *Kornas* 1789 (K).
North Africa, Kenya, Tanzania extending to Asia; also southern Europe. Parasitic on herbaceous plants of river banks and swampy places.

INDEX TO BOTANICAL NAMES